Managing AI in the Enterprise

Succeeding with AI Projects and MLOps to Build Sustainable AI Organizations

Klaus Haller

Apress®

Managing AI in the Enterprise: Succeeding with AI Projects and MLOps to Build Sustainable AI Organizations

Klaus Haller
Zurich, Switzerland

ISBN-13 (pbk): 978-1-4842-7823-9   ISBN-13 (electronic): 978-1-4842-7824-6
https://doi.org/10.1007/978-1-4842-7824-6

Managing Director, Apress Media LLC: Welmoed Spahr
Acquisitions Editor: Jonathan Gennick
Development Editor: Laura Berendson
Coordinating Editor: Jill Balzano

Cover image designed by Freepik (www.freepik.com)

Distributed to the book trade worldwide by Springer Science+Business Media LLC, 1 New York Plaza, Suite 4600, New York, NY 10004. Phone 1-800-SPRINGER, fax (201) 348-4505, e-mail orders-ny@springer-sbm.com, or visit www.springeronline.com. Apress Media, LLC is a California LLC and the sole member (owner) is Springer Science + Business Media Finance Inc (SSBM Finance Inc). SSBM Finance Inc is a **Delaware** corporation.

For information on translations, please e-mail booktranslations@springernature.com; for reprint, paperback, or audio rights, please e-mail bookpermissions@springernature.com.

Apress titles may be purchased in bulk for academic, corporate, or promotional use. eBook versions and licenses are also available for most titles. For more information, reference our Print and eBook Bulk Sales web page at http://www.apress.com/bulk-sales.

Any source code or other supplementary material referenced by the author in this book is available to readers on GitHub via the book's product page, located at www.apress.com/9781484278239. For more detailed information, please visit http://www.apress.com/source-code.

Printed on acid-free paper

To my family.

Table of Contents

About the Author

Klaus Haller is a senior IT architect and IT project manager with more than 15 years of experience in the IT industry. Originally from Germany, he has called Zurich in Switzerland home for many years. He currently works as a senior security architect for a global insurance company, focusing on protecting public cloud infrastructures and data management and AI environments. Klaus is passionate about designing complex solutions that fit into corporate application landscapes. He understands the interplay between technology, operations, engineering, and the business from his previous experience in various roles such as software engineer, project and product manager, business analyst, process engineer, and solutions architect. His expertise includes core banking systems and credit applications, databases, data analytics and artificial intelligence, data migration, public cloud, IT security, and IT risk management. He loves the outdoors and enjoys writing for magazines and online blogs and speaking at conferences and seminars.

About the Technical Reviewer

Robert Stackowiak is an independent consultant, adjunct instructor at Loyola University of Chicago, and author. He formerly was a Data and Artificial Intelligence Architect at the Microsoft Technology Center in Chicago and, prior to that, worked at Oracle for 20 years and led teams supporting North America focused on data warehousing and Big Data. Bob has also spoken at numerous industry conferences internationally. His books include *Design Thinking in Software AI Projects* (Apress), *Azure Internet of Things Revealed* (Apress), *Remaining Relevant in Your Tech Career: When Change Is the Only Constant* (Apress), *Architecting the Industrial Internet* (Packt Publishing), *Big Data and the Internet of Things: Enterprise Architecture for a New Age* (Apress), *Oracle Big Data Handbook* (Oracle Press), *Oracle Essentials* (O'Reilly Media), and *Professional Oracle Programming* (Wiley Publishing). You can follow him on Twitter at @rstackow and/or read his articles and posts on LinkedIn.

Acknowledgments

I want to thank my family for their ongoing support. I could not have written this book without the possibility of spending many evenings and weekends preparing my material and without being able to retreat for some weeks when finalizing the book.

My editors, Jonathan Gennick and Jill Balzano from Apress, were crucial for turning my vision of a book into the concrete book you hold right now in your hands. I appreciated Jonathan's focus on getting an easy-to-read and helpful book. Both showed great flexibility when I delivered my book chapters earlier than planned.

I want to thank Robert Stackowiak for reviewing my book. He provided detailed input on where and how to improve the content and helped solve terminology issues.

The book builds on vast knowledge from business and academic experts on IT project and operations management, databases, information systems, and AI. However, I could only compile this book thanks to my previous and current colleagues, project and line managers, supervisors, and customers over the last two decades. They assigned me to various projects helping me to get a broad understanding of IT organizations and concrete technologies. Over the years, I benefited as well from their experiences and from our discussions, giving me fresh and alternative perspectives. They were vital for enhancing my methodology repertoire, too.

A special thanks also to Max Smolaks from AI Business. He published various articles from me, which eventually became the foundation for multiple chapters. Without his encouragement, I might not have written this book.

Introduction

Artificial intelligence (AI) is what the old west was in the 19th century and the moon in the 1960s: the new frontier. It inspires entrepreneurs, business strategists, and is the hope of CEOs, CFOs, or COOs. It lures ambitious engineers and experienced managers wanting to be part of this big revolution. They all share a passion – the passion for finding their unique role and contributing their experience to help AI initiatives succeed. This book paves the road for data scientists, CIOs and senior managers, agile leaders, and project or program managers to successfully run artificial intelligence initiatives. The focus is neither on the glamorous, glittery world of corporate business strategies nor on the world of math. It is on how to connect these two worlds. The book helps you manage an AI Delivery Organization and AI Operations (Machine Learning/ML Ops) with highly specialized data scientists such that your organization provides the AI capabilities the top management and business strategists need for transforming the organization. It helps to master the project and technology management and the organizational challenges of AI.

The challenge is, against popular belief, not finding qualified data scientists. Universities do a great job educating them. The challenge is to form a team and an organization that delivers artificial intelligence solutions in a corporate setting, to integrate the solutions in the IT application landscape, and to run and maintain them. The field is young, so there is a shortage of managers and operations engineers in artificial intelligence with five or ten years of corporate experience. Today, ambitious IT professionals have an excellent chance to fill any of these roles. If you want to deepen your know-how about enterprise AI, that is, about all the hard work it takes to make AI work in an organization, this book is your key to the new frontier of artificial intelligence in the enterprise.

It condenses experience from more than 15 years in the financial industries and with IT service providers, including roles as a project manager, solutions architect, consultant for core banking systems, process engineer, product manager for analytics, and information systems in general. It is the outcome of more than one year of structuring thoughts and writing them down, many evenings, early mornings, and weekends plus a few months of full-time research. The result, the book you read right now, provides concrete know-how for the following areas (Figure 1):

- Looking from a strategic perspective, how do organizations benefit from AI? What is the essence of a data-driven company – and how does AI help? (Chapter 1)

- How do you successfully deliver AI projects, covering everything from scoping to AI models and integrating them into the overall IT landscape? (Chapter 2)

- Which quality assurance measures help validate the AI models' accuracy and their integration in the overall IT landscape? (Chapter 3)

- How do ethical discussions and laws and regulations impact the work of AI specialists? (Chapter 4)

- How can you move from projects to a stable service delivery organization and a successful team? (Chapter 5)

- What are the technical options to store data and deliver the data to the training environments for data scientists or the production systems? (Chapter 6)

- How can organizations protect data, information, AI models, and system environments? These are all valuable assets. Securing them is essential. (Chapter 7)

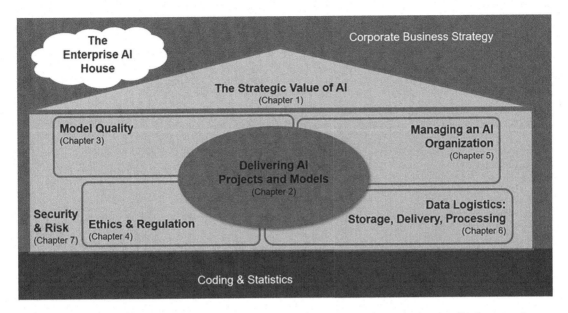

Figure 1. *The Enterprise AI House Structuring the World of AI – And This Book*

Rather than discussing all existing theoretical ideas, academic literature, various schools of thought, statistical or mathematical algorithms, and piling footnotes and references, this book compiles and presents concepts and best practices, guiding managers through the most critical decisions. It helps to structure and to organize the work and to guarantee that your initiative stays on track. The aim is always to help readers excel in their careers by providing information they do not get easily elsewhere.

At the end of this introduction, I wish you an inspirational read. Congratulations on starting your journey to the new frontier of enterprise AI. I look forward to your feedback on how your organization and you as a person succeeded and grew further based on the know-how I share with you in this book.

CHAPTER 1

Why Organizations Invest in AI

In some companies, managers seem to have a pact with AI service and software providers to treat AI as magic: No one understands what happens, but AI promises to deliver exactly what everyone hopes for, even if nobody knows what the result should be. Data scientists love such projects. They can do whatever excites them most – at least until some higher-level manager pulls the plug and terminates the project.

Such project failures root in mixing two aspects: first, what gets managers interested and, second, which projects deliver value to the company. Getting managers interested is simple because they have no other choice. The *Harvard Business Review* ran several articles about AI and data-driven organizations, catapulting this topic to the board level. CEOs and COOs want and have to transform their organization. The challenge is the second part: shaping and delivering a successful AI project. Here, the CIO's role is to build and run the technical foundation, for example, AI and data management capabilities. The excellent news for IT is that funding is guaranteed under such circumstances. Instead, they face a different challenge: identifying the right software products, services, and partners in a fragmented market. It feels like an AI gold rush at this very moment. Corporations invest and set up programs; startups work hard on innovations; consulting companies readjust their business. Vendors rebrand old-fashioned services to sound sexy and appealing in the new world.

A complex market, new requirements, and the pressure to meet the board's expectations are a daunting combination for senior managers. They face a situation similar to a giant puzzle with hundreds or thousands of pieces. Randomly selecting a puzzle piece and finding matching ones tends to be less successful than investing some

1

K. Haller, *Managing AI in the Enterprise*, https://doi.org/10.1007/978-1-4842-7824-6_1

time to get an overview first. Consequently, this chapter puts AI into the corporate context of a data-driven organization by answering the following questions:

- What are the specifics of data-driven companies, and how does AI help?

- How can projects and initiatives calculate the business value of AI projects and services? How can project managers craft a business case?

- How do data-driven companies innovate with AI? Typical sales use cases, and innovative ideas from the fashion industry illustrate how the business perceives AI.

- Why might it be a bad idea to broaden the scope of existing data warehousing (DWH) and business intelligence (BI) teams to cover also AI services?

The Role of AI in Data-Driven Companies

Artificial intelligence and data-driven organizations – these are probably the two hottest topics for most corporations at the moment. What is unique for each of them, what are the commonalities? C-level managers and corporate strategists need answers when shaping innovation and transformation programs.

In short, the term "data-driven organization" covers two aspects – first, the more technical need for collecting data systematically throughout the entire company. Second, the organizational shift in decision-making, focusing (more) on facts and less on seniority. Artificial intelligence helps to make smarter decisions by generating insights and identifying correlations in data humans cannot find in a manual analysis.

In our daily life, we use terms such as data, information, knowledge, or insights interchangeably. Everybody understands, based on the context, what is actually meant. However, precise definitions ease the understanding of data-driven organizations and the role of AI (Figure 1-1).

Data is the raw form: 1s and 0s on disk, as part of a stream from an IoT device, or strings and numbers in databases. On this level, we know nothing more about the data. Data itself does not carry any meaning. It might or even might not help to understand the physical or digital world.

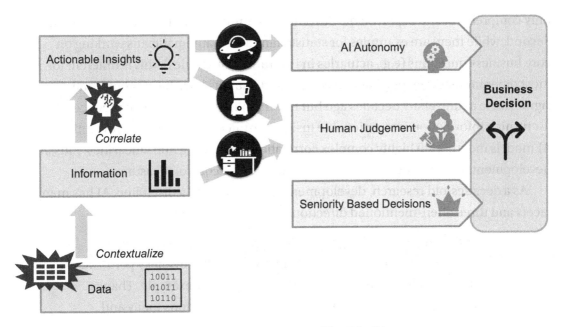

Figure 1-1. *AI and Data-Driven Companies – The Big Picture*

Information is contextualized data. A stream of 1s and 0s becomes a video signal from a camera overseeing an assembly line. In a database, a string now represents a customer name or order ID. A number becomes the profit per year, another the turnover last month. Data becomes information – and has meaning now.

In the foreseeable future, contextualization is and remains a predominantly human task. Automating the identification of employee or customer names or order IDs is straightforward. Understanding whether a table stores order IDs of open orders or of orders where a customer complained is much more difficult to automate. Only humans understand complex data model semantics. They can intellectually penetrate all subtleties of similar-looking, yet different-purposes serving data elements such as open, delivered, and returned orders.

Actionable insights – also named "knowledge" – represent collections of information that guide decision-making. Are there irregularities on the assembly line? Which customers return orders most frequently? As the term "actionable insights" suggests, someone has to decide whether and which actions to perform – and AI delivers helpful or necessary input and options.

Historically, generating actionable insights is the domain of white-collar office workers. Managers and analysts transform large Excel tables full of numbers into nice-looking diagrams. This traditional approach has two limitations. First, humans can only comprehend and experiment with a limited number of influencing parameters. Second, while there are examples for statisticians and mathematicians working on core business questions (e.g., actuaries in insurance companies), this is different for most other industry sectors. Most managers and analysts have heard about statistical significance at university decades ago but never applied the concept in practice. Thus, for turning information into actionable insights, artificial intelligence is revolutionary. AI models incorporate highly complex correlations, including non-linearities; plus, AI development methodologies emphasize robust quality control measures.

As a decades-old research, development, and engineering discipline, AI has many facets and three often-mentioned directions or schools of thought:

- **Artificial narrow intelligence** isolates or singles out and solves highly specialized tasks that only humans could previously perform: identifying objects in images or playing chess, for example. Thanks to deep neural networks, narrow AI drives innovations today and transforms society and corporations.

- **Artificial general intelligence** aims to build a human-like intelligent system capable of learning, understanding, and behaving like humans and the human brain.

- **Artificial superintelligence** wants to build even better brains that surpass the human brain and intelligence.

Reminiscences of these concepts are already part of our popular culture. Artificial superintelligence was a topic of the movie HER a few years ago. An innovative new mobile phone generation comes with an AI assistant. These AI assistants are intellectually more advanced than humans – and eventually depart from them and their limited intellect. The Turing test, a 1950s concept from academia, is widely known today. For passing the Turing test, an AI system must be indistinguishable from a real human conversation partner in a chat. Is this already general AI? No. Building an AI system that mimics a human reflects only a very limited aspect of general AI. Also, the Turing test does not require any learning capabilities.

Today's commercially successful chatbots do not even try to mimic human conversation styles as much as possible. They make it clear that they are just bots. They are not designed to pass a Turing test or achieve the level of general AI. They are perfect examples of narrow AI. They simplify buying or booking processes for online customers or querying support databases by providing a more human-friendly interface than, for example, an SQL command-line interface.

Narrow AI improves the creation of actionable insights. Based on these insides, decisions require some form of judgment. Should a bank clerk call ten potentially interested customers if he can expect that eight might sign a contract? Should he better call 30 instead of ten clients to make 11 instead of eight sales? There is an intruder in the building. Do we call the police, or should the night security guard check the situation?

The rise of narrow AI changes corporate decision-making. An AI model beats specialists with 20 years of experience and a senior title. In the past, there was often no other option than seniority-based decision-making, and most experts have years of experience in their field. They do not just command in dictator-style to prove their power. However, they have no chance against a well-designed AI model, though AI is not the end of senior experts. They remain essential for companies, just with a new focus: for deciding in areas without a dedicated AI model, for challenging whether specific AI models benefit the organization, and for final decisions based on proposals from the AI.

In general, three decision-making patterns are worth mentioning:

- **Decisions based on summarized information**: A domain specialist gathers all relevant data, looks at correlations and dependencies. He decides himself or prepares a proposal for a management decision.

- **Combining the power of AI with human judgment**: AI generates the insights, the final decision stays with humans. This pattern brings benefits for one-time strategic decisions, such as whether to enter the Russian or better the Australian market next year.

- **Autonomous algorithms**: The AI model generates proposals and directly triggers their execution – without any human intervention. This pattern fits operational routine decisions, especially when human judgment is too expensive or too slow. Placing online ads is the perfect example. Algorithms must choose a suitable ad within milliseconds. Plus, a single ad generates only low revenues – a few pennies or sometimes Euros. Human intervention is too expensive and too slow for such use cases.

When considering how data becomes information and information actionable insights and how the different decision patterns work, the idea of being data-driven and the role of AI get evident in the big picture (Figure 1-2). Data-driven organizations require data and, thus, invest in collecting data and its contextualization to derive information. It is the base for deciding on facts. Either humans or algorithms and AI turn the gathered information into actionable insights, AI is beneficial for repetitive decisions and for filtering out relevant parameters from large and complex data sources. Plus, AI models decide in (near) real-time once the model is trained and do not induce high costs for each decision.

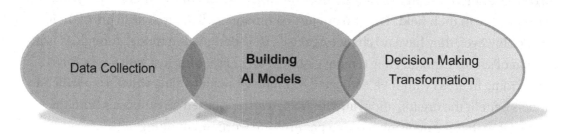

Figure 1-2. *The Three Challenges for Data-Driven Organizations Using AI Capabilities*

When companies become data-driven and introduce AI, their business cases must not cover only the AI capabilities. They must incorporate the expenses for becoming data-driven: establishing capabilities for collecting data from data warehouses or business applications, from IoT devices, and from systems in engineering or service departments – plus expenses (and time and resources) for reengineering the decision-making processes (Figure 1-2). Looking at these three challenges holistically together – data collection, building AI models, transforming corporate decision-making – is the key to success.

Calculating the Business Value of AI

A C-level commitment to become a data-driven company and to invest in AI is a nurturing environment for AI projects. Still, building and establishing a concrete AI service needs funding, and funding requires a convincing business case. The business case can be a hurdle for highly innovative projects because quantifying their benefits is a challenge. Luckily, economic research identified four ways IT investments generate value. They help to craft a concrete business case. These four ways or options are:

- **Informational benefits**, for example, delivering information to the decision-makers.

- **Transactional benefits** such as improving daily operation and cutting costs.

- **Strategic benefits**, for example, changing how the company competes in the market or its products' nature.

- **Transformational benefits** refer to future benefits, for example, due to additional features of a new tool that employees start using once they get acquainted with the tool, and that enable them to optimize their work.

There is a lot of anecdotal evidence for benefits but little systematic empirical work covering many companies. One exception is a study Elisabetta Raguseo performed on French companies of various sizes. In her research, she looked at Big Data technologies. They handle and analyze massive data volumes coming in at high speed and in varying formats. Big data relate closely to data-driven companies or the guiding aim of AI to generate insights from large amounts of data, making her study relevant for this book, too.

Elisabetta found out that enhancing employee productivity is the most relevant transactional benefit in all considered industry sectors. Reducing operating costs was second. The subsequent most relevant strategic benefits are aligning IT with a business strategy and enabling a quicker response to change.

In the area of transformational benefits, expanding capabilities is rated as most important. It can be achieved, for example, by enabling new employees to start quickly working on integrating additional data sources. In the area of informational benefits, they list proving management data, improving data accuracy, providing data in more useable formats, enabling easier access to data, and enabling faster access to data.

There are two key points to consider when applying the insights for your business case for AI projects. First, the area is close, but not the same. Raguseo's results are good inspirations, not god's word. Second, company culture and politics impact which of the **benefit types** work best in your organization and your managers:

- Certain benefits are **easy to quantify** – especially (but not only) transactional benefits. The project calculates today's costs and estimates the potential savings after implementing the project. Such business cases are perfect for companies driven by financial numbers with a strong influence of controllers and like-minded managers.

- There are **vague to quantify** benefits. Will predictive analytics increase sales by 1% or 10%? The calculation in the business case and the actual costs and savings can differ widely. Many strategic and transformational benefits fall into this category. Such arguments convince visionary and innovation-driven decision-makers, not nitpickers.

- The informational benefits are the easiest to fund. When a manager believes to need data, she will fund a project or service if within her budget. There is **no need to quantify** any benefits beforehand.

When the business value is clear, the management decides whether to invest in the transformation to become a data-driven organization and build and extend its AI capabilities. In most cases, AI initiatives start with the low-hanging fruits with easy-to-measure results such as improving and optimizing sales.

Use Case: Sales Operations Efficiency

Improving sales efficiencies is one example for realizing transactional benefits: more sales, less effort. The use case takes advantage of the power of data (and algorithms) to boost sales and to identify customers potentially terminating their contract. In such use cases, the AI predicts, for example, which customers might buy a product.

Traditionally, sales agents or the sales management would filter customer lists based on their gut feeling to get an idea which customers they could call. It was and is their responsibility to make the right choices when selecting the customers to call. They are responsible for the outcome.

For an AI-generated list, the data scientists are (at least partially) responsible for the sales success. The lists are clear guidelines for sales agents on how to act. If you want to sell office furniture, these are the customers that most likely buy. Call them! These customers might cancel their cable subscriptions. Do something and prevent it! Whoever creates such lists is responsible for their quality.

How and Why Up-Selling, Cross-Selling, Churn Use Cases Thrive

There are convincing reasons why cross-selling and churn AI use cases are widespread. First, they are straightforward to implement. AI projects can often rely on the data of a single source system such as an ERP system, a CRM system, or a data warehouse for creating a good model. The advantage: AI projects do not have to bring data from various sources together, which is often a time-consuming task or even a project risk. Second, the transactional benefits are apparent and easy to validate. The AI project goes in the ring against a seasoned marketing and sales professional. Both the marketing professional and the AI choose 50 customers each. These 50 each are the customers, the marketing professional, respectively, the AI thinks they are the customers most likely buying the product. Then, sales reps contact these one hundred customers. They try to close a deal without knowing about the experiment. In the end, you count how many customers bought from the human-generated list, how many from the AI-provided one. The superiority of the AI list should be striking.

This use case, however, needs the active support of the sales organization. What happens if the sales staff opposes the idea? Luckily, data scientists can implement a different use case just with data, without requiring help from other teams: the churn prediction use case. A churn prediction AI model predicts, for example, which customers cancel their mobile phone plan subscription soon. The data scientists generate the list – and wait. Maybe a week, perhaps a month. Then, they check whether their prediction became a reality and present to their management how many customers the company lost because the business ignores AI-generated insights.

Upselling and churn prediction models base on scoring customers, usually assigning values to them between zero and one to customers. The scores indicate how likely a customer buys a product or cancels a contract. Organizations prioritize their call-center resources to contact customers with a high score. They call a customer with a score of 0.57 before reaching out to customers with a score of 0.12 (Figure 1-3). The input variable columns in the table reflect what we know about the customers. Based on these input variables, the scoring function, aka the AI model, calculates the customer affinity for a golden credit card.

Customer	Input Variables for Scoring					Customer Affinity for the Specific Product	
	Age	Income	ClientAdvisorID	Bundle	Funds	ScoreCCGold	
Igor	25	3'000	12	Yes	1	0.43	
Ling	31	5'000	12	Yes	3	0.57	
Louis	59	9'000	12	No	3	0.61	
Camille	54	3'500	23	Yes	2	0.12	
Nathan	19	1'200	95	No	0	0.05	
Laura	61	13'000	53	No	2	0.30	

Figure 1-3. Customer Scoring, a Typical AI Use Case

The effect of using the scores for prioritization is similar to champagne bubbles. "Interesting" customers get to the top of the list as champagne bubbles rise to the top in a champagne flute. Figure 1-4 illustrates this, with the left representing contacting customers randomly and, on the right, choosing them based on a score.

Suppose a call center agent calls a potential customer from this list. If he is interested, he buys because of the call. If we call a customer that is not interested, we "lose" one call.

The table to the right in Figure 1-4 contains the same names; just the order changed, thanks to AI. More customers who want to buy the product are at the top of the list and get called first. In this example, the sales agents close only one deal after four calls for the left table. In contrast, three of the four approached customers sign with the score-based approach reflected in the table to the right. This increase in sales results from the power of AI and analytics – and managers can implement the same and calculate the transactional benefit for their use cases.

	Classical sales				Sales with the power of AI and predictive analytics				
Customer	Buys if contacted?	Contacted?	Sold?		Customer	Score	Buys if contacted?	Contacted?	Sold?
Igor	No	Yes	No		Nathan	0.83	Yes	Yes	Yes
Ling	Yes	Yes	Yes		Laura	0.77	Yes	Yes	Yes
Louis	No	Yes	No		Yelena	0.53	Yes	Yes	Yes
Camille	No	Yes	No		Igor	0.43	No	Yes	No
Nathan	Yes				Ling	0.42	Yes		
Laura	Yes				Paul	0.37	No		
Paul	No	No	No		Maria	0.33	No	No	No
Maria	No				Camille	0.21	No		
Yelena	Yes				Louis	0.20	No		

Figure 1-4. How AI Supports Sales If You Have Limited Resources for Calling Potential Customers. Using AI, You Can Increase the Chance to Contact Interested Customers

Hurdles and Prerequisites for Success

Enthusiastic AI evangelists, software vendors, or external consultants sometimes forget to mention two strict **limitations or prerequisites** for the applicability of AI for sales. First, AI requires a sufficiently large training data set. AI algorithms generalize from this training data and learn and create models. No (good) training data means no (good) model.

There is no *historical training data* when launching a new product the first time, for example, an eco-friendly gold credit card with blinking pink LEDs. Nobody bought this product already. Thus, there is no training data. Creating an AI model is impossible.

The second strict limitation is that these AI use cases help only for **resource shortages**, for example, if the sales organization cannot contact all potential customers. Resource shortages are common in the retail banking mass market, but not in wealth management or when a big European airplane manufacturer sells airplanes to airlines. In the latter cases, each customer generates high revenues. The sales staff's costs are low compared to the revenues just one deal generates. Airplane manufacturers can afford large sales teams and over-capacity. The implications for AI: no shortage means no need to invest in AI. The sales team is in constant contact with their customers on the C-level and knows who might buy 20 airplanes next year without AI. Potential customers might even approach the airplane manufacturer anyhow. There is no shortage in sales capacity per customer AI could optimize. The bottleneck is how long the sales team can talk with airlines' top management. CEOs do not have the time to talk hours and hours with one supplier every week. Thus, AI can help identify hidden needs and cross-selling options, enabling the sales staff to present the right ideas and products. In other words: What are the five products and services we should tell a specific customer when we visit him next week? It is an AI functionality everybody knows from online shops: "customers who buy steak and crisps buy German beer as well."

Besides these two strict limitations – need for adequate training data and AI being helpful (only) in case of resource shortages requiring optimization – there is a third hurdle relating to the business model and company culture. Certain industries and business models rely on a close **personal relationship** between sales staff and customers. Examples are the insurance sector or private banking for super-wealthy customers. Here, the sales staff might see AI as a direct attack on their professional pride, self-conception, and value as an employee. AI (and the IT department) claims to understand customers better than their bank advisors, who know their wealthy clients for decades and met them in-person. Furthermore, the sales staff's compensation depends on closing deals – and their skill to sell the right product to the right customers. If computers do better, client advisors fear for their long-term relevance for the organization – and their bonuses and commissions.

Fear for their relevance is a work culture aspect that exemplifies that data scientists alone cannot transform the organization into a data-driven company relying on AI for more efficiency. A data scientist can and must create a good model. Dealing with the skepticism, for example, of sales staff, is a change management task. Their managers and the organization's top management have to handle this topic and decide. They can stop the use of AI in their area – this happens in companies. However, innovative managers push forward anyhow. They even ignore furious sales staff or navigate and circumvent the issues, for example, by initiating a direct online sales channel.

Organizational Impact

How does AI impact employees in the sales organization if AI takes over scoring customers in and for sales and marketing? First, assigning scores is nothing new. Marketing and sales professionals have been doing this for decades. AI revolutionizes **how to calculate scores and who** executes the task. It does not base anymore on the gut feeling of a marketing professional, who, based on 30 years of experience, really, really knows that men between 35 and 55 are the ideal customers. The modern form of scoring bases on AI, incorporating as much information as possible such as age, gender, buying history, or clicks on products and ads in the webshop.

When AI takes over the scoring, this impacts the **sales management** organization. The IT department and its software solutions create the lead lists of whom to call to sell a particular product from now on. The sales management can shift its focus away from playing around in Excel to higher-value tasks such as designing campaigns and understanding to whom to sell brand-new products and services.

The impact on the **sales staff** depends on the detailed business model. Is there a deep personal relationship? Does the (time-consuming) personal relationship translate into adequate extra revenues? Do the sales reps consider customers to be "their" customers and is this belief backed by reality? For example, in (outbound) call-centers, AI-generated lead lists do not change how agents work. As before, they get a list of (potential) clients they have to contact. The only difference is that they should close more deals because the customers on the AI-generated list contain more customers likely to buy. Thus, they generate higher revenues (or reach the previous goals in less time). Sales staff with a **personal relationship** will see some change because AI complements their personal understanding of their customer with statistically derived insights. Especially in the latter case, the success depends on a combination of IT and data science and an organizational transformation.

Insights for Product Strategies

AI brings not only quick wins for sales organizations on a tactical or operational level. It also provides strategic insights, for example, for product and service managers. We exemplify the latter aspect of generating strategic insights by looking at the benefits of AI for product management and sales and at the methodology behind it.

Identifying the Business Challenge

Increasing revenues and profits is part of the DNA of every successful company. Product managers are eager to find new ways to boost sales for their portfolio. The previous section already presented one specific example of how to push sales. Here, we take a broader, more holistic perspective: the Ansoff Matrix (Figure 1-5, left). The Ansoff Matrix covers the best-known growth strategies to increase sales. Market penetration is the most straightforward strategy to implement: selling more of the existing products to the existing customer base – the approach presented in the previous section about operational efficiency (Figure 1-5, middle).

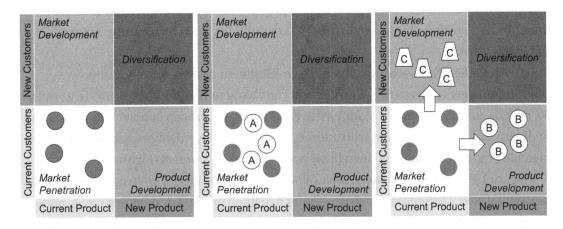

Figure 1-5. *Ansoff Matrix with the Illustration of the Various Sales Strategies. The Diversification Case Is Not Illustrated Due to Its High Failure Risk and the Less Apparent Benefits from AI*

Product development is another growth strategy: selling new products to an existing customer base. For example, many years ago, banks launched credit cards and sold them to their existing customer base. It is not realistic to get such product ideas from AI. However, it can help matching customer segments and products to identify niches.

This approach does not apply an AI model to predict who signs up for a credit card. This strategy requires **analyzing the AI model itself**. Potential insights from the model can be:

- Teenagers do not have a credit card.

- Adults in low-income households seldom have one.

- Retired persons often have no credit card. If they have one, they use it infrequently.

These insights help product managers to launch product innovations. In this example, the bank has unserved customer segments, for example, teenagers to whom the bank must not grant loans and adults with high default risk. The product manager needs a new product because the existing offering is not suitable for these customer segments. We all know the solution: prepaid credit cards. It looks like a credit card, but you can only spend as much money as you put on the card beforehand. For retired persons skeptical about the risks when paying with a credit card, an assurance for cyber fraud is an option. Thus, AI can help with its model to shape and develop new products (Figure 1-5, right/B). Another outcome of such an analysis can be that the customer base of similar products overlaps. In such a case, product managers might want to restructure their product portfolio because the products cannibalize each other and each product has fixed- costs.

An AI model helps as well for a market-development growth strategy. Suppose a product manager analyzes the AI model and can characterize potential buyers (e.g., over 30, no kids, no cat, potentially a dog). With this knowledge, he can contact market research or direct marketing companies with address databases. They might provide lists of persons matching the criteria, making them likely potential new customers (Figure 1-5, right/C).

Figure 1-6 illustrates the mechanics of how to use an AI model for insights from a procedural side. The idea is to check whether a defined product or market strategy and observed customer behavior match. If not, this means there is an opportunity for growth when readjusting the product and market strategy and checking again.

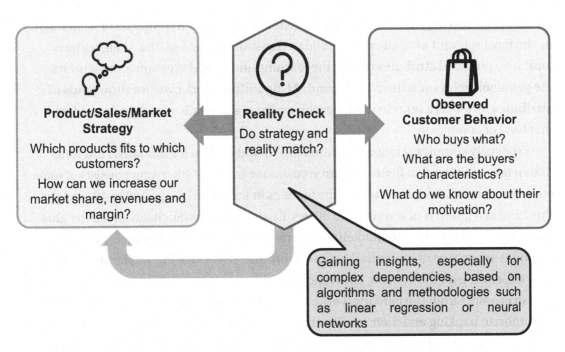

Product/Sales/Market Strategy

Which products fits to which customers?

How can we increase our market share, revenues and margin?

Reality Check

Do strategy and reality match?

Observed Customer Behavior

Who buys what?

What are the buyers' characteristics?

What do we know about their motivation?

Gaining insights, especially for complex dependencies, based on algorithms and methodologies such as linear regression or neural networks

Figure 1-6. *AI for Strategic Product Management*

AI and Analytics vs. Excel

When AI project managers or data scientists explain to another team the wonderful new world of AI that revolutionizes this other team's work, they must never forget one truth: the rest of the organization worked well before the AI professionals got involved. Thus, AI professionals might have to explain over and over again why AI-generated lists outperform manual Excel-based approaches. I remember vividly a statement from a bank's marketing specialist who told me: an apprentice generates a prospects list within two to three days with Excel. They do not need AI.

When looking at product management, comparing sales strategy and sales reality is nothing new. However, especially analytical or statistical methods allow for an enhanced understanding compared to the traditional manual Excel approach. **Excel** works for 20 or 30 criteria. It is suitable for manually validating the impact of one or two of these criteria such as gender or age, or both. Such simple hypotheses for manual validation could be "Only customers under 25 order energy drinks" or "Families with children like to buy pink bed linen with unicorns and flamingos."

15

AI brings two improvements. First, there is **no need for a hypothesis** of what might be the most relevant attributes to be validated manually. Instead, the AI algorithms look at all provided attributes during the learning and model creation and identifies the genuinely relevant attributes. Second, AI algorithms work even for **thousands of attributes**, whereas a human working with an Excel sheet fails when confronted with that level of complexity.

Especially complex relationships are hard to understand for humans. How long does it take with Excel to figure out three customer groups? One group consists of male customers under 25, one of the wealthy females in the suburbs, aged 30 to 40, and the third and last group is of seniors over 60 in villages with dogs. Such advanced insights help product managers and marketing experts to identify and model their **personas**. Personas are groups of (potential) buyers or product users sharing similar demographics and, depending on the context, motivations.

An example from the banking world: A bank wants to understand why customers in corporate banking end their relationship. They realize: The bank advisors sold a newly launched product quite well to their commercial clients. That is the knowledge generated using analytics. Drawing the conclusion, judging, and deciding is the task of the bankers. They have to figure out why customers left after buying a well-selling product. They learned that the bank advisors pushed the product onto their clients whether it was suitable or not. Customers felt tricked and terminated their banking relationship. This churn use case example illustrates how data-driven companies benefit because they get insights no one could have asked for. And those insights help companies move ahead of their competitors. Nobody saw the relationship between the well-sold product and customer running away – the AI model made it obvious.

Understanding the AI Project Deliverables and Methodology

How do data scientists and AI projects generate strategic insights for product managers, as discussed in the previous section? This challenge has many facets. The starting point is always real-world data serving as training to create an AI model. For the example of a sales scenario in Figure 1-7, this is the table "Sales Data used for Model Training." The more data, attributes, or parameters, the better. Crucial is the column that contains the expected output or prediction. In the example, this is the "Yes/No" column. Who might buy – this is the ultimate prediction the model should make.

The data scientists apply learning algorithms to this training data, aiming for creating a predictive model. The model is the base for strategic and tactical insights. Tactical insights are customers that might have a high probability of buying the product. To this aim, data scientists apply the predictive model to customers who have not been contacted by sales reps yet. They get a score for each customer – this is table "Lead List" – of how likely they will buy the product if approached. The score helps to prioritize whom to call.

The data scientist looks at the model for getting strategic insights. The AI model contains the information that middle-income customers have a strong affiliation with the product. Other indicators for potentially product-affine customers are being male, having a high income, or paying for a banking bundle.

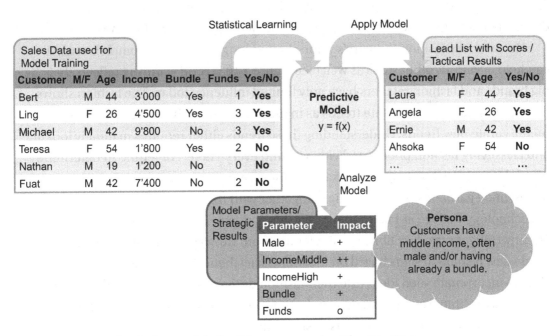

Figure 1-7. *AI for Sales and Sales Management – The Big Picture*

AI-Driven Innovation in Fashion

The typical AI bread-and-butter business is identifying upselling or cross-selling opportunities and understanding customer characteristics. However, AI can be cool and trendy – or at least data scientists can help others become and stay fashionable and stylish. The fashion and especially the fast-fashion business seems to be just about finding cheap factories for manufacturing in the Far East and spending big on marketing campaigns. However, AI is crucial in fashion and penetrated many critical functions than, for example, in the highly digitized world of banking.

The massive usage of AI comes as no surprise when asking yourself whether you would put the same designs and the exact same sizes on stock in the stores in North Dakota and your flagship stores in Milan and Paris. Do your customers have the same fashion sense everywhere? Should you price your items the same around the globe? The questions are provocative – at least if you live in North Dakota – but H&M lost much money when shipping similar items to all its stores. They failed to sell enough clothes at undiscounted prices as planned, which significantly impacted their revenues and margin. Thus, H&M invested in "localization." AI predicts which store's customers are most interested in which fashion piece and route their garments accordingly. The predictions base on point-of-sales data and information about their loyalty club members (and where they shop).

However, the fashion industry's value chain is longer and more complex, and AI can help in various other steps as well (see Figure 1-8). **Trend scouting** reflects that big brands and fashion retailers have only limited influence today. The fashion shows in Paris and Italy do not dictate trends as in the 20th century. Influencers and social media curate new trends. Understanding these trends either requires manually looking and classifying fashion photos and videos found everywhere – or letting AI determine trends by categorizing images, for example, for garment style or color. When it comes to designing garments, Tommy Hilfiger, together with IBM and the Fashion Institute of Technology College in Manhattan, run a joint project. They used AI technology, Tommy Hilfiger's product library, and pictures from runways to enable students to create new fashion designs based on the brand's aesthetic. It is unclear whether they want to incorporate such ideas into their fashion design methodology. Still, the example illustrates one important point: data and AI can also innovate **creative processes**. Similarly, H&M sold a Data Dress. Its design is the result of the person's life and context, such as the weather. These influencing parameters determine the choice of color, material, and additional details to make a really customer-specific dress.

Figure 1-8. *Fast-Fashion and AI*

Next in the value chain is estimating the sales success of products, and **ordering** the matching amount of pieces is a typical prediction question for AI. Once it is clear which items to assign to which store, the next step is pricing the pieces. Which price are customers willing to pay? How many pieces do we estimate to sell for which price? H&M reports benefits of using AI – and especially of using a combination of AI and final human judgment.

Finally, there are client-facing functionalities such as apps, emails, or push messages. They help implement the already mentioned use cases such as upselling and cross-selling or identifying shopper groups and characteristics. While this is standard across industries, the AI-based Tommy Hilfiger **brand and shopping experience** is an exciting high-tech idea: "see-now-buy-now." Customers can buy fashion directly during a fashion show when models present the garments on the runway. Similarly, an app allows customers to identify fashion pieces on camera snapshots and provides the opportunity to buy them directly in the online shop. It is an inspirational eye-opener even for non-fashionados to see in how many ways AI innovates the fashion industry. It should motivate business executives and data scientists in other industry sectors to look around with open eyes to identify innovative opportunities in their daily work as well.

Business Intelligence and AI

All the previous examples about sales and product management or fast fashion exemplified how AI innovates companies. When managers want to start a similar program, the first question is: Who will run the project? Do we have a qualified team with relevant experience?

Most companies have data warehouse (DWH) and business intelligence (BI) teams. These teams know the company data very well. They run a platform with at least all necessary technical features for the first AI use cases. However, I was first surprised when I noticed that many companies – especially medium-sized organizations – establish separate AI programs and data science teams instead of enhancing the DWH or BI team with data science know-how.

This phenomenon highlights: just having the best and brightest experts, even if they and their software solutions match the business needs, is not enough. Organizational compatibility is essential. The following four reasons prove that. They illustrate why companies establish AI teams separately in addition to BI and DWH teams:

1. BI teams often focus on summary **reports**, not on generating additional knowledge and insights (see the discussion about data, information, and knowledge reflected in Figure 1-1). Not every reporting-focused BI team can reinvent itself (and has the funding for that).

2. The **sponsors** differ. Often, the funding for a data warehouse comes from controlling or marketing and sales. Thus, the teams have to deliver results valuable for their sponsors. In contrast to data science teams, they cannot shift many engineers to projects outside their sponsor's interest area.

3. Data warehousing teams often focus on developing thoroughly tested and well-specified solutions. This cautious approach is necessary, for example, for regulatory reports. The expectations in AI differ. Suppose the business needs **quick answers** for a one-time-only question that is not asked again. Not all data warehousing teams are ready for that.

4. Small data warehousing teams are often busy keeping the systems and all feeds up and running. They have to ensure a smooth daily, weekly, or monthly delivery of business-critical reports. Such teams are strong in **operations** and engineering but miss know-how for more advanced and mathematically challenging AI tasks.

These are four aspects organizations should consider before assigning AI-related responsibilities to an existing DWH or BI team. However, as we see in later chapters, a data warehouse is an excellent data source for data scientists. An existing DWH speeds up many AI projects. Thus, it is essential for an organization that BI and DWH teams and data scientists understand that they complement each other. They do not compete.

Summary

At the end of this first chapter, I want to summarize the main points quickly:

- Data-driven companies base their decisions on facts and data rather than on experience and seniority. AI helps in extracting actionable insights from large and complex data sets.

- The four benefits of AI are informational (providing additional information), transactional (making something better or cheaper), strategic (impacting a company's or product's standing on the market), and transformational (paving the road for future improvements).

- AI is about creating models. These models have a tactical benefit for the organization. They can predict individual behavior, for example, whether a person is likely to buy a phone. The models provide strategic insights, too, such as identifying the characteristics of typical buyers.

- AI helps in optimizing single tasks such as cross- and upselling. Alternatively, AI can enable transformations of business models, such as is illustrated with the fashion industry example.

- Many companies have DWH and BI teams that could take over at least more straightforward AI projects. However, separate AI teams are crucial when service models, funding, and governance structures differ.

With this knowledge, we are ready to look at how to run AI projects and deliver innovation. Just be warned and remember the old proverb: "We admire what we do not understand." After the next chapter, you understand how data scientists work and how you can improve projects. So, AI will be less magical for you afterward!

CHAPTER 2

Structuring and Delivering AI Projects

Congratulations, the board of directors wants to move forward with AI. The senior managers chose you as their AI program and project manager. Now it is your time to deliver results. You had the vision; now you can make your vision become a reality. Just do not forget: some truths apply to all projects, AI or Non-AI ones. The corporate world expects projects to deliver results cost-efficiently, timely, and reliably. Most companies cannot and do not want to pay ten PhDs and wait for eight months to get a perfect solution. Academia is patient and invests so much time and effort. Companies, not so much. As an AI project manager, your challenge is to work on a solution with two or three IT professionals and deliver the first results in six or eight weeks. With such a context in mind, this chapter provides answers for the following AI project management challenges (see Figure 2-1):

- What are the different layers on the AI technology stack which enable companies to innovate with AI?

- How can AI project managers scope their projects to meet the expectations of their business stakeholders and sponsors?

- AI models are the core deliverable of AI projects. How do, for example, neural network or statistics-based AI models look like?

- What is a development methodology which works for AI projects? What are vital efficiency factors?

- How can you integrate AI models in an existing corporate application landscape?

© Klaus Haller 2022
K. Haller, *Managing AI in the Enterprise*, https://doi.org/10.1007/978-1-4842-7824-6_2

Figure 2-1. *Structure of the Chapter*

The Four Layers of Innovation

Managing AI projects and teams has some peculiarities. Team members tend to be more technology-affine and have a stronger computer science and math background but less experience with business problems than legacy application developers. There are fewer patterns and established toolchains known from traditional software engineering. Thus, AI project managers have to provide more directions to ensure delivering concrete business benefits. They have to challenge their data scientists constantly to prevent the latter from doing AI research and reinventing the wheel.

Structuring a technology area in layers and defining a technology stack is a long-known and widely practiced approach. When looking at the example of Java, the Java backend application runs code in Java VMs. A Java VM runs on a hypervisor, which runs on an operating system, with the operating system running on physical hardware. There are similar layers in AI (Figure 2-2). The lowest layer is the **hardware layer**. Neural networks, for example, require many, many matrix multiplications. Regular CPUs can perform them, but there are faster hardware options: Graphical Processing Units (GPUs), Field Programmable Gate Arrays (FPGAs), or Application Specific Integrated Circuits (ASICs). Innovation on this layer comes from hardware companies such as AMD or NVIDIA. Companies buying such innovative products are server and workstation manufacturers or public cloud providers. These customer groups need hardware for their data centers that run AI-rich workloads efficiently.

The **AI frameworks layer** forms the next layer. It represents frameworks for machine learning and AI, such as TensorFlow. TensorFlow distributes the workload for training massive neural networks on large and heterogeneous server farms, thereby taking advantage of their computing power. TensorFlow abstracts for the data scientists from the underlying hardware. It parallelizes autonomously computing tasks. Data scientists can run the same code and learning algorithms on a single laptop or a cluster with hundreds of nodes with hardware tuned for AI and benefit from the cluster. Data scientists do not have to change their code for different hardware configurations, saving them a lot of work and effort.

Companies use AI frameworks to develop new AI algorithms and innovative neural network designs. Suppose a company wants to develop the next big thing after GPT-3 or a fundamentally new computer vision algorithm. In that case, their data scientists use an AI framework to engineer and test new algorithms. In other words: nearly all data scientists do not innovate on this layer but work on top of the existing frameworks.

There are only two scenarios for innovation in this layer: improving an existing framework such as TensorFlow or developing a completely new one. These activities are usually part of academic research. For companies, it makes sense only if they have an outreach to a massive number of data scientists. The latter applies to public cloud providers or AI software companies. They need innovation to provide "the best" AI framework to lure data scientists into using their AI and/or cloud platform.

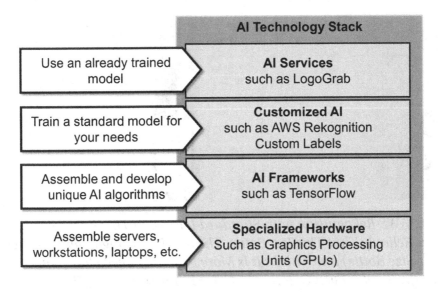

Figure 2-2. The AI Technology Stack

Next is the customized AI layer. Its specifics are easy to understand after the purpose of the top layer is clear, the **AI services layer**. AI services enable software engineers to incorporate AI into their solution by invoking ready-to-use AI functionality. The engineers do not need any AI or data science know-how. One example is Visua/LogoGrab. The service detects brand logos in images or video footage. It measures the success of marketing campaigns, for example, by checking how often and for how long the brand of a sponsor is on TV during a sports event. It finds counterfeits and look-alikes of brand products on the Internet and electronic marketplaces as well. LogoGrab is an example of a highly specialized AI service.

There are also ready-to-use AI services with a broader focus, for example, from AI vendors or the big cloud providers Microsoft Azure, Google Cloud, and Amazon AWS. Examples are AWS Rekognition or Amazon Comprehend Medical. The latter service analyzes patient information and extracts, for example, data about the patient herself and her medical condition. AWS Rekognition (Figure 2-3) supports use cases such as identifying particular generic objects on pictures such as a car or a bottle. Do you see more customers drinking beer, white wine, or red wine? Marketing specialists can feed festival pictures to the AWS service, let the service judge what is on the image, and prepare statistics showing the results. Knowing that 80% of the festival visitors drink wine, 15% champagne, and just 5% beer might be a red flag for a brewery when deciding whether to sponsor this festival.

Figure 2-3. *AWS Rekognition with Standard Labels. In This Example, the Service Determines Reliably That This Is a Bottle. The Service Struggles with the Left One (It Is a Half-Size Bottle) and Thinks It Is More Probably Beer Than Wine. On the Right Side, the Service Even Detects That This Is a Red Wine Bottle*

Companies can innovate quicker when building on existing AI services such as LogoGrab. They integrate such a service into their processes and software solutions. They do not need an AI project or data scientists. There is no risk of project delays or failure due to issues with the AI model.

In many cases, generic AI services are not sufficient for concrete scenarios. What should a marketing specialist do if she needs to understand whether there is Swiss wine on a picture, other wine, or other drinks? She needs a neural network tailored and trained for her specific niche.

As stated, AWS Rekognition comes with a set of standard object types the service can identify, for example, on pictures. However, AWS Rekognition is more powerful. Engineers can train their specific neural networks to detect object types specific to their needs. The engineers have to provide sample pictures, for example, for Swiss wine bottles. Then, **AWS Rekognition Custom Label** trains a machine learning model for these customer-specific object classes.

This AWS service is just one example of services forming the **Customized AI layer**. They train and provide ready-to-use customer-specific neural networks based on customer-delivered customer-specific training data. In Figure 2-4, a marketing specialist for Swiss wine might be interested in understanding whether festival visitors prefer Swiss wine, other wine, or other drinks. So, she prepares training and test data with pictures labeled for these three types of drinks. When pushing the "train model" button, AWS generates the neural network without any further input and without requiring any AI knowledge.

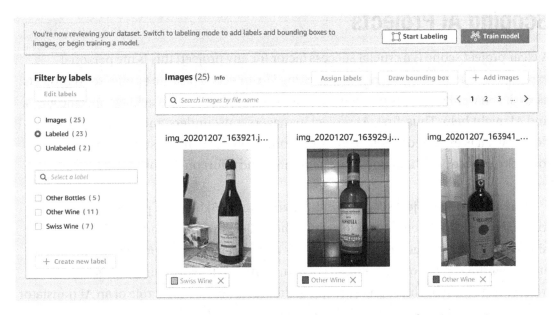

Figure 2-4. *Preparing Training and Test Data Sets in AWS Rekognition Custom Label (Console View). Data Scientists Can Upload Pictures and Label Them*

Customized AI is intriguing for companies that want to optimize specific business process steps to gain a competitive advantage. Thanks to customized AI, they achieve such goals without a large data science team. They can use cameras to check products on the assembly line for flaws in the production process. Therefore, they collect sample pictures for products that can and cannot be shipped to customers, let the customized AI service train a neural network, and apply this AI model to images coming from a camera on top of the assembly line. All typical suspects offer customized AI services: cloud providers plus analytics and AI software vendors.

AI Services, Customized AI, AI Frameworks, and AI-specific hardware – AI innovation comes in many forms. For innovation, companies and organizations rely primarily on university-trained data scientists. These data scientists know the AI frameworks layer exceptionally well. However, this layer is more for academic researchers but not in the focus of (most) companies. Thus, managers have to communicate their strategy clearly: on which layer(s) is the company innovating with AI? AI Services, Customized AI, AI Frameworks, or Specialized Hardware? This four-layer AI technology stack can act as a communication tool in any AI project and for any AI manager.

Scoping AI Projects

A clear project scope is a crucial success factor for any project; this is my personal experience from managing and restructuring IT projects. This rule applies to AI projects as well. Senior managers fund them because they want or have to achieve specific goals, and AI might help. Thus, first, AI project managers must understand these goals and deliver what is expected. Their second supporting goal is to make sure the project team focuses on these goals and does not spend time and effort on optional or irrelevant topics.

The six-step AI project scoping guidelines (Figure 2-5) helps AI project managers to reach this goal. Depending on the setup, the project manager works on these topics himself (potentially together with a solutions architect and an AI specialist). Alternative and depending on the organizational context, he might delegate these topics to the traditional roles of business analysts and architects or to the newer role of an AI translator that we discuss in a later chapter. What matters are a good understanding of business topics and AI technologies combined with social skills. Once the project has completed these six steps, the technical implementation starts, that is, working on the model and integrating the overall application landscape, hopefully following a structured, methodological approach.

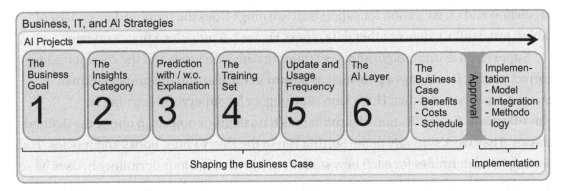

Figure 2-5. Six Steps to Scope an AI Project – and Shape a Business Case

Understanding the Business Goal

The first step is a rather generic project management task: understanding what the business wants to achieve. It is a clear sign of trust in AI's innovation potential when the business spends half a million or three million euros on implementing an AI solution. Then, they obviously believe that AI helps them run their business better. However, to prevent disappointments, a business case must answer the following questions:

- What is the exact goal of the business? Why do they invest money?

- What are the criteria to decide whether the project delivered everything they expected?

- How does the project relate to strategic and tactical business goals?

- What is the expected timeline?

- Is there already a budget? How much is it?

The answers to these questions help when writing the actual business case. The AI project manager can verify whether the expectations of the management and how the actual high-level project planning after the scoping phase match. Do the project's direction, deliverables, and timeline match the senior management's expectations?

Understand the Insights Category

The second scoping step translates the business problem into an AI problem. Is supervised or unsupervised learning the solution? Which subcategory is needed? Clustering, association, or dimensionality reduction for unsupervised learning or prediction and classification for supervised learning? Does the training data consist of tables, text, audio, video, or other data types? Figure 2-6 provides a first overview.

Supervised learning algorithms get training data with input and the corresponding correct output. It is always a pair such as a word in English and the proper German translation: <red, rot> <hat, Hut>. One subcategory for supervised learning is **classification**. A classification algorithm puts data items or objects in one of the defined classes. They work similarly to the Sorting Hat in the Harry Potter books and movies. The Sorting Hat determines for each new student in which of the four dormitory houses he or she fits best in and should live in the following years. A typical classification use case is image recognition: Is it a cat in the picture, or is it a dog?

Prediction is the second category of supervised learning insights. How much sugar and cream should we order to produce enough ice cream tomorrow? What will be tomorrow's ice cream sales based on last year's numbers, today's weather, and the weather forecast for tomorrow? Supervised learning is often directly actionable. Thus, they are beneficial for transforming operational business processes in companies and organizations aiming to become data-driven and AI-empowered.

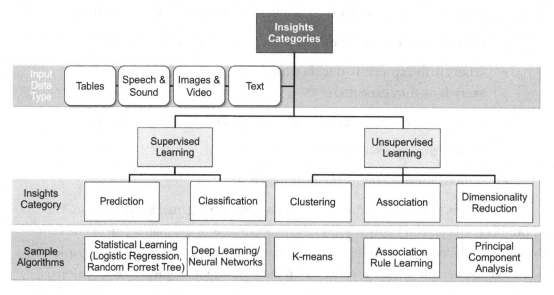

Figure 2-6. *Insights Categories – A Quick Overview*

Unsupervised learning structures the input data. A **clustering** algorithm looks at all data points. It comes up with, for example, three groups of data points that are pretty close together: status-oriented customers, price-sensitive customers, and nouveau-riche customers. Clustering provides additional insights to the sales and marketing departments and product managers without directly triggering any actions. **Associations** identify relationships between data elements, whereas **dimension reduction** simplifies complex, high dimensional data, for example, to remove noise from images.

In general, unsupervised learning algorithms unveil hidden structures of/in your data, but (often) do not directly tell what to do. While the book uses mainly supervised learning algorithms as examples, many of the methodologies work for unsupervised learning algorithms as well.

When a project has identified the needed insights category, the next step is to understand the data types used for training the model – and where to source them. **Database or CSV/Excel tables** are common formats, though the exact format or file type does not matter much. What matters is that the machine learning algorithms get structured, table-liken input.

Images and videos are other types of input data. They help with camera images during the production process and check whether the produced items have a good quality or video streams from CCTVs to check for persons entering restricted areas. **Speech and sound** (e.g., engine noise) or **text** in the form of documents or emails are additional options.

The exact data types do not impact whether AI can come up with the solution per se. It is more to make a better effort estimation (in short: tables are less work than speech) and elaborate the timeline more precisely. Furthermore, the AI project manager can validate how experienced the data scientists are with the particular area of AI needed for this business question. No project manager wants to realize at halftime that his project misses essential skills, that it is unclear whether suitable AI code libraries exist, or that they must identify a particular external service to build the solution.

Having answers for the questions addressed in the second scoping step brings the project closer to deciding which AI methods and algorithms to use. However, this decision still needs clarity regarding the explainability question.

"Prediction Only" vs. "Prediction and Explanation"

Two similar questions require slightly different approaches and potentially different AI learning algorithms. Suppose a TV streaming platform's sales department wants to boost sales. First, they could ask which customers they should call to sell as many subscription upgrades as possible within one week. Second, they might be interested to understand how their customer base differs from the overall population. For example, the product manager could realize her customers are mainly in the German-speaking part of Switzerland. Adding a French and an Italian language offering might be an option to foster growth in Switzerland.

The first option with lists of potential buyers is operational. The sales managers need a list of which clients to call or mail. They do not care whether the list comes from a statistical model, a 100-layer neural network nobody understands, or from a fortune-teller using a crystal ball. They want a list with accurate predictions of who might buy.

The second question asks about buyers' characteristics. Again, data scientists need a model that predicts who is likely to buy what. This time, though, understanding and explaining the model is essential. The attributes and their values matter. They help to distinguish buying customers from others. A fortune-teller or a 100-layer neural network might be highly accurate with predictions, but less accurate statistical models are often better for understanding characteristics. "Explainability" is a prominent topic in one of the later chapters.

Consequently, AI project managers or data scientists can provide a "prediction only" AI model or must deliver a "prediction and explanation" one. It is like painting a house white or pink. Both are possible. You should just clarify very clearly what your customer expects to prevent surprised reactions later.

The Training Set Challenge

Training AI models require good training data. The better the training data, the better the trained model's accuracy. The AI training algorithm is undoubtedly essential, but – as a rule of thumb – an AI model cannot be better than the quality of its input training data. Five top biologists discussing which plant on a particular picture can prepare better training data than a drunken monkey doing the same. Suppose the idea is to be "better than a human" in the decision-making. There are three explanations when and how this can happen:

- Imitating a better-than-average human expert ("five top biologists").

- Automatically derived real-world training data, for example, from shopping baskets or server logs.

- Humans get tired and, thus, make more mistakes, AI components don't.

So, automating human judgment, for example, in quality assurance or incident analysis, requires a knowledgeable person to prepare the training set.

For supervised learning, the training data depends on sample input values plus the expected, correct output. Training a model for distinguishing between tomatoes and cherries requires many sample images with a label stating whether a concrete image is a cherry or a tomato. Getting such labeled training data is a challenge for many projects.

In general, there are four main approaches: logs, manual labeling, downloading existing training sets from the web, and database exports. **Existing training sets** enable researchers to develop better algorithms for common problems such as image classification. They often help to validate algorithms for challenges for which many solutions exist. Thus, in general, they benefit more academia. Companies should check whether already (pre-trained and) ready-to-use models exist. Not surprisingly, commercial AI projects often use **logs** (e.g., the shopping history of customers or click logs showing how customers navigate a web page) or time-consuming **manual data labeling** (e.g., humans mark whether an image contains a cherry or a tomato) for building training data sets. Building training sets from **database and data warehouse exports** works often for AI scenarios in the business or commercial domain such as sales, potential in combination with logs.

Model Update and Usage Frequency

A significant cost block for AI projects is the **integration** of AI components in the overall application landscape. It requires considerable time and engineering resources to make AI components and existing and new applications work together smoothly. Closely related is **automating**, collecting training data from various source systems and cleansing and transforming the data to create an AI model.

AI project managers must understand to which extent their projects need work-intensive integration and automation functionality. Key decision parameters are model usage and model update frequency. The **model usage frequency** indicates how often the AI model, for example, classifies pictures or predicts customer behavior. Is the model for a one-time, very specific mailing campaign and never used again? Is it for a webshop that calculates daily shopping items to present prominently? The more often the model is used, the more beneficial is integrating the AI model into the application landscape. Integration reduces manual work by automating the transfer of files and data from AI training environments to production systems. This technical integration minimizes the operational risks of manual data transfers, such as copying wrong files or forgetting to upload new data. The practical implications of such mistakes can be severe. A webshop might propose items not matching customer needs. Male teenage customers might not spend more money with a webshop that tries to sell them expensive pink handbags. Thus, a frequently used model calls for a well-tested integration of the AI components with other applications to reduce manual work and operational risks.

Closely related to model usage frequency is the **model update frequency**. It refers to how often data scientists retrain the AI model. Retraining means keeping the model structure, the data sources, the output data structure, etc., and "just" update the model's parameter values so that they reflect the latest data. The retraining frequency depends on how often the real world and its reflection in data changes. When a sports fashion shop trained a model for the winter and skiing season, the shop should not use the same model next summer when it is super hot, and everyone spends the weekends and evenings on the beach or at outdoor pools. Suggesting pink skis as an add-on to a pink bikini might get the webshop in the press but not generate extra revenues. We observe a similar pattern as for the usage frequency: if data scientists have to perform the same process – this time, the one for updating the model – frequently, it is time to automate collecting the data and preparing it for retraining the model.

The conclusion for AI project managers is simple. During their project scoping phase, they must understand how deep the integration of the AI model and the AI training environment has to be with the rest of the IT application landscape. The answer heavily influences project costs and timelines.

Identify the Suitable AI Layer

The last scoping question aims to clarify how much the AI project can build on existing external AI solutions and services. This topic relates to our discussion of the AI technology stack at the beginning of this chapter. A project can build on ready-to-use AI Services. In such a case, the project engineers a solution with AI functionality without requiring any AI skills in the project team. Then, there is the area of customizing pre-trained AI models. The pre-trained model, for example, detects cars, but the project wants it to distinguish a BMW from a Citröen. Finally, AI projects can use, for example, TensorFlow and architecture neural networks best suited to their needs.

This last scoping question about the AI layer requires more AI knowledge than the previous ones. A project might even have to research the market to understand whether suitable external services or tools provide the envisioned AI functionality. These questions have a massive impact on the project and its business case. They determine whether and how many data scientists the project needs.

From Scoping Questions to Business Cases

The scoping questions help AI project managers to clarify many details before inviting senior data scientists and software architects (or external vendors) to work on a high-level project plan.

A project plan at this stage identifies the main milestones and potential valuable intermediate deliverables such as a minimal viable product. It states when which milestones can be roughly achieved and lists the needed resources. Resources cover how many engineers with specific skills are necessary, potentially highlighting internal candidates for critical roles. Resources cover as well the financials for external staff, licenses, and computing infrastructure, whether internal servers, powerful laptops, or public cloud resources.

Preparing a project plan and the needed resources is standard for project managers for non-AI and AI projects. Often, they have to follow company-specific mandatory rules and processes and fill out standard templates.

A business case just covering a project plan and expenses would be incomplete. The management wants to understand the benefits and the potential return of investment as well. The previous chapter covered potential benefits of AI projects already in more detail.

Eventually, the big day comes. A senior manager or a management committee decides whether you, as an AI project manager, get the funding for the project based on your business case. You maximize your chances with a clear scope, a convincing high-level project plan, and target-specific elaborated business benefits. If you get the funding, the data scientists are happy to prepare for and start working on the envisioned AI model, their core project deliverable.

Understanding AI Models

AI models aim to solve particular, narrow problems. A model might get CCTV footage as input and provide as output whether a "moving thing" on the camera is a harmless cat or a potential burglar. Over more than half a century, research and engineering in AI came up with various approaches to generate such insights. They fall into two main categories: computational intelligence and symbolic artificial intelligence.

Solutions based on **symbolic artificial intelligence** represent knowledge explicitly, for example, using first-order logic or a formal language. The systems have formal operations or rules that operate on their symbolic knowledge. Such systems were popular when I went to university in the late 1990s. Thus, some readers might be familiar with the names of some of the concepts and languages in this area, such as state-space, means-end analysis, blocks world, expert systems, or Prolog and Lisp.

An easy-to-understand example is the following Prolog code fragment. The system knows some facts about who likes whom in a small group:

```
/* Let's start with some facts ... */
likes(peter, john) /* Peter likes John*/
likes(peter, andrea) /* Peter likes Andrea*/
likes(peter, sophie) /* Peter likes Sophie*/
likes(john, peter) /*John likes Peter*/
```

Interference rules generate new knowledge and drive the system's reasoning. In this example, a rule defines being friends as two persons who reciprocally like each other.

```
/* A rule to define friendship ... */
friends(x, y) :- likes (x,y), likes (y,x)
```

Posing a query to Prolog starts the reasoning process. The system tries to prove or disprove a statement (Are Andrea and John friends?) or determine values for which the query is correct (Who are Peter's friends?).

```
/* Let's pose queries to let the system reason a bit ... */
?- friends(john, andrea)
No
?- friends(peter, X)
X = john
```

Today, the second big group of AI methods, **computational intelligence**, is much more popular. It approaches knowledge and facts representation and reasoning and insights generation entirely differently. Computational intelligence methods store information or facts as numbers. There is no explicit knowledge representation. Insights generation means multiplying and manipulating big matrixes.

Figure 2-7 provides an example. The input picture is a big matrix with numbers representing color and brightness. The system then implements matrix operations – the reasoning and insights generation. The figure contains the very first matrix operation, which averages cell values to reduce the matrix size. After some more matrix operations, the system states whether the image is a car, truck, boat, or plane.

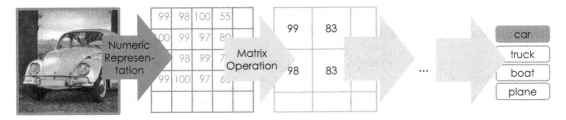

Figure 2-7. *Computational Intelligence: Image Content Representation by Numbers, Further Processing Also Based on Numerical Operations, in This Case by Calculating the Input as the First Processing and Reasoning Step. (Picture source:* `https://pixabay.com/photos/vw-beetle-oldtimer-classic-1396111/`*)*

Computational intelligence algorithms are at the heart of the ongoing AI revolution in enterprises. While data scientists know which algorithm and method to use in which context, AI project managers need a basic understanding to contribute to the "definition of done" or quality assurance discussions. Readers interested in the multitude of algorithms find them covered in detail in one of the various algorithmic-centric AI textbooks.

We focus in the following on two examples: an old-fashioned statistics method (some might not even consider it to be AI) and deep learning neural networks. For the latter, we present basic neural network concepts and discuss advanced topologies. These algorithms require some entry-level math, though understanding the essence is possible even without.

Statistics-Based Models

Mathematicians and statisticians developed various algorithms over the last decades and even centuries. Many work perfectly for table-based Excel-like input data, typically available in ERP or core banking systems. We focus on one algorithm and just look from a high level: linear regression. Suppose an online marketplace for secondhand cars wants to incorporate a "fair price" feature. It should make deals smoother by fostering trust between car dealers and customers by suggesting a "fair price" to the contract partners. Therefore, the marketplace needs algorithms to estimate, for example, the market price for a Porsche 911 3.8 Turbo Cabrio.

In the first step, the model might consider only one input parameter for price estimations: the car's age. Since we use linear regression, the mathematical function to determine the car price is a linear function:

$$Price\left(Age \ of \ the \ Car\right) = a_1 * age + a_0$$

Once we know the price of two car sales – for example, 153,900 EUR for a 12-month-old car and 124,150 Euro for a 28-month-old one – we put these two data points in a diagram and draw a line through them. The line is a linear function for our car-price estimation (Figure 2-8, left).

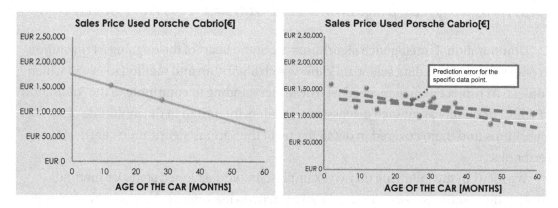

Figure 2-8. *Using Linear Regression for Price Predictions*

However, a good pricing model needs hundreds or thousands of data points. Figure 2-8 (right) illustrates this. The example contains many data points, meaning the model accuracy should be good. However, it is (nearly) impossible to draw a line going through all data points if you have more than two. So, how can linear regression help? Basically, there are two needs:

- An error metric measuring how good a function reflects reality. A metric makes two or more estimation functions comparable. We can measure which of them estimates sales prices better. A widely used metric for that purpose is "least squares" (see Figure 2-9).

- An algorithm that systematically or randomly tries out various values for parameters such as a_0 and a_1 in the example. Using the error metric, it calculates the quality of a specific parameter setting. It readjusts the parameters and calculates the error again – until the algorithm decides to terminate the optimization and returns the best values for a_0 and a_1.

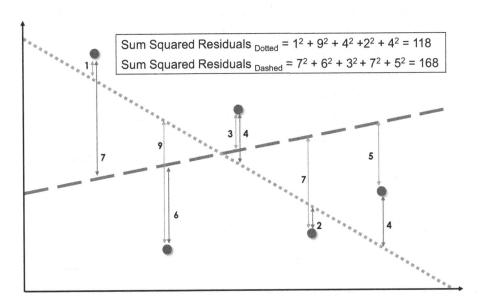

Sum Squared Residuals $_{Dotted}$ = $1^2 + 9^2 + 4^2 + 2^2 + 4^2$ = 118
Sum Squared Residuals $_{Dashed}$ = $7^2 + 6^2 + 3^2 + 7^2 + 5^2$ = 168

Figure 2-9. Least Squares Illustrated with Two Estimation Functions (Dotted and Dashed Lines). The Algorithm Calculates for Each Data Point the Derivation Between Prediction Value and Actual Data (Residual). Then, the Algorithm Squares the Residuals and Sums Them Up. The Result Is 118 for the Dotted and 168 for the Dashed Line. Thus, the Dotted Line Is the Better of the Two. However, an Even Better One Might Exist and Requires Additional Experimentation

We already mentioned two impact factors for model quality: the learning algorithm and the size and quality of the training data set. Similarly important are which attributes the model incorporates. In our example, the fair-price estimation function for cars considers just one attribute: the age of a vehicle (v_1). The function is insufficient even if trained with millions of sales data points. The reason: various other factors impact the price of a car as well significantly, for example, the number of driven kilometers (v_2), whether the vehicle was involved in a heavy accident (v_3), has air conditioning (v_4), or a high-end music system (v_5). As a result, we get a more complex price estimation function to optimize:

$$Price(v_5, v_4, v_3, v_2, v_1) = a_5 * v_5 + a_4 * v_4 + a_3 * v_3 + a_2 * v_2 + a_1 * v_1 + a_0$$

The underlying assumption for such a linear regression function is that the nature of the problem is linear. However, used car price estimation functions are **not linear**. The value of a car drops heavily when a car dealer sells it the first time to a customer. The price decline slows down afterward, and, usually, the sales price does not get negative. Depending on the problem's nature, replacing the linear estimation function with a different (more complex) mathematical function results in better estimations.

Another algorithm similar to linear regression is **logistic regression**. Logistic regression helps for classification tasks such as deciding for specific credit card payments whether they are fraudulent. Usually, a function such as the sigmoid one is applied in the end to provide a classification value between 0 and 1 (see Figure 2-10, left). If the value is 0.5 or higher, it might be a yes, otherwise a no.

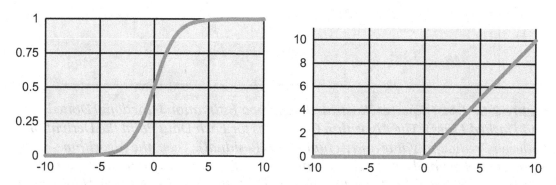

Figure 2-10. *Sigmoid Function (left) and Rectifier – or ReLU, Rectified Linear Unit – Function (left)*

Neural Networks

Neural networks serve a similar purpose as the statistical regression algorithms. They allow for more complex and accurate prediction or classification logic. Thus, they are especially beneficial for tasks such as image recognition and natural language processing.

Neural networks date back to the 1950s when Rosenblatt developed the Perceptron for detecting patterns in (small) images. Key algorithms are from the 1980s and 1990s. Their big breakthrough was in the 2010s resulting in their widespread use in industry today. Enabling factors have been the explosion of computational power and available data such as text, images, audio, or video. The cloud is the most recent catalysator for the adoption of neural networks. Cloud providers offer highly scalable computing resources and ready-to-use AI services.

The human brain, with its neurons and synapses, is the inspiration for artificial neural networks. Neurons are the processing units. Synapses connect neurons to exchange information, which drives the information processing of the neural network. A neuron fires and is activated, depending (also) on the activation status of the incoming neurons.

An artificial neural network implements the **reasoning** of a single neuron in two steps. First, the neuron calculates the weighted sum of all incoming activations from the feedings neurons. In Figure 2-11, the neurons A, B, C, and D propagate their activations a_A, a_B, a_C, and a_D to neuron N. Neuron N calculates the weighted sum of these activations using the weights w_{AN}, w_{BN}, w_{CN}, and w_{DN}. So, the calculation is: $a_A*w_{AN} + a_B*w_{BN} + a_C*w_{CN} + a_D*w_{DN}$.

When the weights are positive, a connection supports that neuron N fires. If the weight is negative, this suppresses the firing of neuron N. Besides the already mentioned weights, a value named "bias" is typically added to the calculation, resulting in the following term: $a_A*w_{AN} + a_B*w_{BN} + a_C*w_{CN} + a_D*w_{DN} + b_N$.

The bias shifts the graph on the x-axes. The purpose of this becomes clear when looking at the second processing step of the neuron: the neuron applies an activation function on the weighted sum, resulting in the activation of this neuron. A widely used activation function is the Rectifier (ReLU) function (see Figure 2-10, right). It returns the weighted sum of the activation sum if positive; otherwise, 0. This activation is then fed to other neurons, such as neurons X, Y, and Z in Figure 2-11.

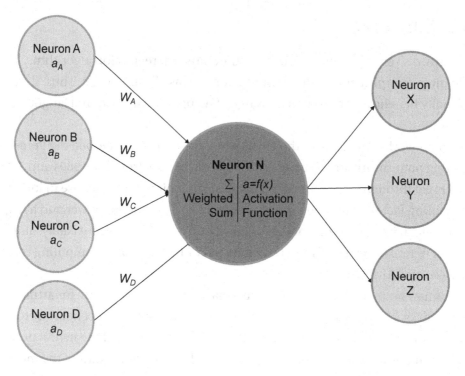

Figure 2-11. *Model of a Neuron*

Neural networks consist of large numbers of neurons grouped into layers (Figure 2-12). The first layer is the input layer. Its neurons represent the outside world, such as pressure sensors' values in a turbine or pixels of a camera. The last layer is the **output layer**, which delivers the artificial neural network's reasoning. It consists of one or more neurons, depending on the neural network's purpose. For binary classification problems – does an image contain either a dog or a cat – one neuron is sufficient. A value close to 1 indicates a dog, a value close to 0 a cat. Suppose a neural network classifies animal pictures, for example, whether the image is a dog, a cat, a dolphin, or a manta ray. In that case, the output layer consists of various neurons for all the different animals.

The neural network can have zero, one, or multiple **hidden layers** between the input and the output layers. In the most straightforward neural network topology, all neurons of one layer feed their activation to all neurons of the following layer – a fully connected neural network.

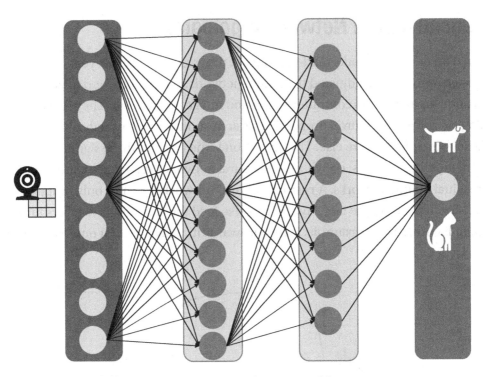

Figure 2-12. *Neural Network Topology with Two Hidden Layers*

When **training a neural network**, the first step is to choose a data item from the training set as input. We calculate the neural network's prediction or classification for this item and compare the result of the neural network with the correct output. We know the correct result since the data item is from our training set. The difference between the correct and the actual result is the input for step 2, which readjusts the weights. For this readjustment or learning step, data scientists commonly use the backpropagation – a gradient-descent – algorithm. The backpropagation algorithm starts with optimizing the output layer's weights and then works its way back layer by layer.

The weights are parameters of the neural network architecture representing the brain or intelligence of the artificial neural network. The parameters that control the learning process are called hyper-parameters. The neural network architecture, such as the number of layers and neurons, is a hyper-parameter, and so is the learning rate. The learning rate influences how much weights are adjusted if the actual and the correct output differ for a training data item. A high learning rate enables the neural network to learn quickly in the beginning. Over time, the learning rate must drop to ensure that the neural network converges to stable weights.

Advanced Neural Network Topologies

Research on neural networks made significant progress over the last years. Data scientists improved and adjusted the neural network architecture for dedicated application areas, such as computer vision and image recognition. They relax the topological rule that a neuron propagates its activation value to exactly all neurons of the next layer and nowhere else. The relaxation is the base for three significant neural network topology variants: convolutional neural networks, recurrent neural networks, and residual neural networks. A **residual neural network** does not only propagate its neurons' activations to the next layer. It can also feed the activations to layers further down. In Figure 2-13, for example, layer 3 forwards its activation not only to layer 4 but also to layer 6.

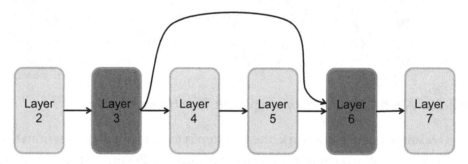

Figure 2-13. *Residual Neural Networks*

Convolutional layers are another topological variant. They differ from fully connected neural network layers in two ways (2-14). First, neurons of a convolutional layer do not consider the activation of all neurons of the previous layer. They consider only specific collocated neurons, such as neighboring pixels of an image. For example, only a 3x3 matrix of neurons in a layer feed their activations as an input of a specific neuron in the next layer. Second, the neurons of a convolutional layer calculate their activation differently. They use a filter (named convolutional kernel) with weights. The filter is a kind of window moving over the matrix, which has the activation values of the previous layer.

In Figure 2-14 (right), the left upper cell of the kernel has the value "1". When calculating the value of the black cell, the left upper value of the activation matrix with the value "0.9" is weighted with the value "1". Then, "0.7" is multiplied with "0" and "-1" with "0.1". This calculation continues for all other cells of the filter matrix. For calculating the value of the grey cell, we start with "0.7", which is multiplied with "1".

Data scientists build on neural network topologies with convolutional layers for dedicated AI application areas such as computer vision with their complex problems. They consist of twenty or thirty layers, a mix of fully connected layers, residual layers, and many convolutional layers.

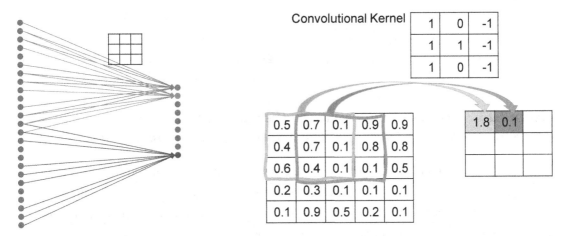

Figure 2-14. *Advanced Architectural Patterns for Neural Networks: A Convolutional Layer/Convolutional Neural Network. The Left Side Illustrates the Topology Aspect, That Is, the Wiring of Neurons in Two Layers, the Right a Sample Filter Matrix*

In time series prediction or natural language and speech processing, **recurrent neural networks** are a popular neural network topology choice. The topology reflects that the ordering of the input is essential. When understanding a text, "Heidi kills Peter" and "Peter kills Heidi" have a different meaning. Recurrent neural networks connect nodes within one layer to reflect the ordering aspect (Figure 2-15). Sophisticated recurrent neural network topologies base on the well-known neural network architectural patterns such as Long Short-Term Memory Units (LSTM) or Gated Recurrent Unit (GRU). They introduce the concept of a "memory" for previous state information. The relevance of memories is easily understood when thinking about crime stories. The book might start with the sentence "Heidi killed Peter." This sentence contains information still relevant two or twenty sentences later when the police arrest Heidi.

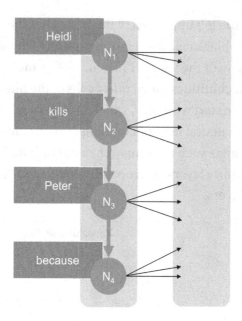

Figure 2-15. *Recurrent Neural Network. The Characteristic Is the Connection Between the Neurons in One Layer Represented by the Bold Green Arrows*

AI project managers benefit from a basic understanding of advanced neural network topologies, even though they do not design or optimize the topology. They should be aware that various best practice topologies exist. Plus, they should be sparring partners for data scientists suggesting designing and optimizing a company-specific and use-case-specific neural network topology (and investing months of work) instead of using best practice topologies.

The fact that best practice topologies exist impacts project teams. First, it is an extra motivation to use existing, well-trained AI models, for example, from cloud providers. Second, suppose projects embark on the journey to design and implement their own neural network. In that case, the project manager should ensure that he has an experienced data scientist in his project team. Ideally, he can convince a senior data scientist with know-how in the concrete application area. Otherwise, he should try to get at least temporary support from such a person.

Developing AI Models

The last decades taught the IT industry and IT professionals that unstructured and uncoordinated engineering or unsystematic experimentation is the short road. Its only drawback: It leads to disaster, not to success. Thus, AI project managers should ensure that the engineering of "his" AI models results from a hopefully creative and systematic engineering process. In the following, we highlight three topics each project manager should know:

- Jupyter Notebooks as the most widely used AI tool

- CRISP-DM (CRoss Industry Standard Process for Data Mining) as a methodology for structuring AI projects from the idea to the delivery

- Approaches for improving the efficiency of AI teams

The Jupyter Notebook Phenomenon

Jupyter Notebooks are a fascinating piece of software. I learned about them in my first deep learning and AI courses. I never heard about them in my two decades in data management, often with hands-on SQL programming. As Google Trends illustrates, the interest in Jupyter notebooks started to rise from around 2014/2015. It almost doubled every year in the beginning and is still increasing today (Figure 2-16). Every manager in AI will hear data scientists mentioning them on a nearly daily basis. If data scientists have issues with their Jupyter Notebooks, this slows down their work. In general, the primary use of Jupyter notebooks is developing Python scripts. However, they are not simply a kind of TOAD-like tool or SQL Developer for Python. They are a new way to work with data.

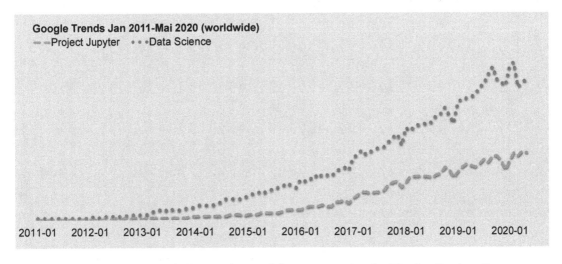

Figure 2-16. *Google Trends Analysis of the Interest in the Topics Project Jupyter and Data Science from January 2011 till May 2020*

Jupyter Notebooks are an **interactive development environment**. The commands support, for example, downloading data and cleansing or transforming the data, and training AI models. As an interactive environment, data scientists can write some lines of code and execute them immediately. They can choose whether to rerun the complete script or just run new lines of code. This choice is essential. Single commands for data preparation or for generating the model can take minutes, hours, or days. Rerunning the complete script for testing some new code lines often takes too long.

Besides the interactive mode, the second differentiator is that **commenting code is fun** with Jupyter Notebooks. It is different from old-style Java code comments and feels more like making notes on a paper notebook. Before (or after) writing a new command, adding a comment helps to remember its exact purpose. Otherwise, data scientists might not know anymore five or ten commands later the reason and actual context of every single line of their code. Jupyter Notebooks allow formatting comments nicely (e.g., bold, italics) and integrating images and sketches. These options indicate that comments are not just an add-on but an integral part of the model development process.

To illustrate the concept of Jupyter Notebooks better, Figure 2-17 contains a screenshot. This Jupyter Notebook is a script for downloading images from the web and storing them in an array for later processing. On the top is a cell with comments ("markdown"). Then, there is a cell with code. The first commands define and instantiate variables, followed by a for-loop for downloading images. The last commands write to the console. Below the script with the code, there is the output the Jupyter notebook wrote to the console when executing this script.

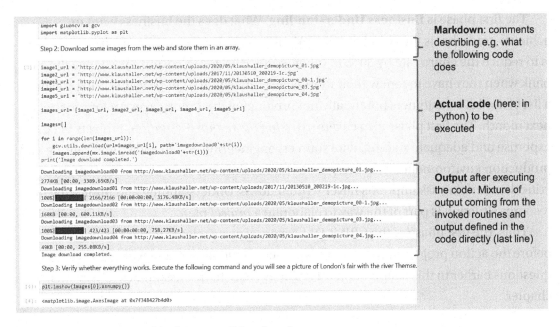

Figure 2-17. *Example of Jupyter Notebook*

CRISP-DM for Model Development

With or without Jupyter notebooks, unstructured experimenting and uncoordinated engineering seldom deliver a successful AI model or a working AI-driven software solution. The bigger and more relevant the AI projects get, the more vital are methodologies. In AI, the most popular one is the Cross-Industry Standard Process for Data Mining (CRISP-DM). Released at the turn of the millennium, it is still state of the art with its six phases (Figure 2-18).

The first phase is **Business Understanding**. What does the business want to achieve – and what is the corresponding AI project goal? An example of a *business goal* is to reduce the churn rate by 50%. In other words: 50% fewer customers change the bank when they have to renew their mortgage. A matching *AI project goal* is to provide a list of the 1000 customers potentially not prolonging their mortgages up for renewal next month. The first phase also covers *assessing the resource situation*. Projects need expertise and adequate staffing, data sources, engineering and AI tools, and training and production environments. There might be legal constraints to be clarified or implicit critical underlying assumptions never written down. Understanding and improving the resource situation is part of the way to elaborate a *project plan.*

Thus, the first phase of the CRISP-DM covers aspects related to the business case before the action project starts. We covered such elements in detail with the scoping questions earlier in this chapter and the business benefit discussion in the previous chapter.

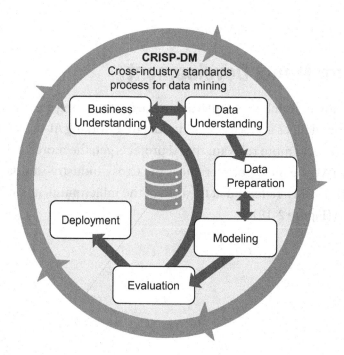

Figure 2-18.   *The CRISP-DM Model*

The second phase is **Data Understanding**. It starts with getting access and potentially copying data from the sources identified in the previous phase. It covers tasks such as getting accounts, opening firewalls, or initiating file transfers. Data scientists describe and work on understanding the data from a high-level perspective. They document the meaning of the attributes, their syntax (e.g., string vs. number fields), their statistical distribution, and their quality. Typical quality issues are, for example, empty fields or tables that contain only partial data (e.g., all customers worldwide, but excluding China or the US).

In phase three – **Data Preparation** – data scientists transform and prepare the final data set for the subsequent model creation. They identify potentially helpful tables and attributes. The rule of thumb is to provide as many attributes as possible and let the AI training algorithms determine which attributes improve the AI model quality. When trying to understand which customers might decide not to prolong their mortgage, everything we know about the customer can help for training the model: account balances, number of contacts with a bank advisor, inflow and outflow of money – everything that widely relates to the customers, its mortgage, or her behavior and actions. Data preparation also covers feature engineering. Not all potentially helpful input values exist in a database. A data scientist might have to extract all customer contacts from client advisors from the company's calendar solution and determine the number of contacts per customer. The net in- or outflow of assets requires summing up the positions today and one year ago.

Data preparation also covers cleansing the data: skipping rows with empty attributes, setting empty fields to a default value, or estimating missing values based on external information. During this phase, data scientists have to bring such data together from various tables and databases. All information about customers must be in one table, requiring merging and joining tables from one or multiple databases. The core database might provide the account balances. The number of meetings with the bank advisor and the customers comes from a table in the customer relationship solution. Finally, sometimes the syntax or format of the data has to be changed: Boolean values 'Y' and 'Y' might have to be mapped to the numbers '0' and '1'.

The fourth CRISP-DM phase is **Modeling**. It is the core task and the one data scientists love most: creating an AI model. They decide whether to train, for example, a neural network or linear regression model. They define how to measure and evaluate the model quality and perform this task after the model creation. Finally, they document the model accuracy and compare potentially competing models.

The **Evaluation** phase leads to the final decision of whether the model is ready for production usage. The stage covers the quality assurance as well as deciding whether the model is good enough for production. In contrast to the modeling phase, which looks at the model from an IT/AI perspective, the evaluation phase challenges whether the model really helps reaching the business goals. It can even mean applying and trying out the model in reality. One bank branch might call customers with a high risk of leaving. After one month, the project compares the numbers for this branch with those of a similar one that did not put in extra effort to convince potentially leaving customers. Such a comparison allows validating the benefit of AI.

The sixth and final phase is **deployment**. The project plans how the model is used, elaborates a monitoring strategy to detect a degenerating model, and ensures the maintenance of the AI. It covers, furthermore, generating a final project report and having a lessons-learned workshop.

Projects usually do not go linearly and in waterfall-style through these phases (see arrows in Figure 2-18). Sometimes, going back one step allows reaching the final goal quicker and/or better. For example, assume an AI project detects that the bank mainly lost German customers last quarter. Then, compliance and sales managers might suddenly remember that they forced them to leave – and did not tell the AI project team. Another iteration becomes necessary. Indeed, the AI project benefits heavily from the data and insights in a second iteration.

In general, in the absence of detailed company-specific data, projects should plan three iterations. A second iteration typically needs around 50% and a third around 25% of the effort of the first. Also, practical experience indicates that the actual creation and training of the AI model takes only 10–20% of the overall time budget. It is a small fraction – and the reason why IT organizations try to improve their data scientists' productivity. They want them to work on their core tasks and on new models that create business value.

Improving the Productivity of Data Scientists

Making data scientists work more efficiently is a big topic, though the exact motivation differs between organizations. Tech companies and start-ups often cannot recruit enough experts with specific technology or domain know-how. Other companies – especially in the IT service or consulting sector – face another dilemma. They (or the managers of their customers) have great ideas about innovating with AI. Most of the time, the data is available. However, my experience from brainstorming workshops is that financial gains and savings and the actual project costs often do not match. No business managers want to invest CHF 100,000 in an AI project to get CHF 30,000 savings per year. AI projects are cool but unfamiliar and, thus, "risky." Hesitating managers might expect bullet-proof business cases. When AI organizations can increase their productivity, costs go down, and the business value remains the same. AI projects pay off quicker and deliver higher savings or returns.

Going through the CRSIP-DM process is a structured way to identify improvement potential (Figure 2-19). The first phase is "business understanding." The broad existing literature about **business analysis** and requirements analysis covers all methodologies and tools for understanding the goals and requirements of users, customers, and managers. Besides this general methodology know-how, there are AI-specific chances to optimize the first phase. This book elaborated them in the **scoping** section at the beginning of this chapter and the business value discussion in the previous chapter.

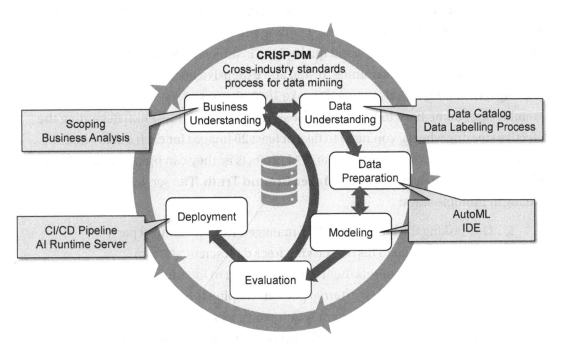

Figure 2-19. *Optimizing Options for AI Projects*

Scoping and business analysis do not reduce the effort data scientists have for cleaning and preparing training data and to train a concrete model. However, a good business understanding reduces the risk that data scientists do not fully understand the business challenge. An insufficient understanding of the expected goals can result – after weeks of intense work – in the verdict that an AI model is correct and well-trained, but does not help anyone.

The second CRISP-DM process phase, "data understanding," covers data collection and acquisition, understanding the actual data, and verifying its quality. Two relevant optimization options exist for this phase: managing training data and setting up or introducing a data catalog. A **data catalog** compiles information about the various data sets stored in the organization's data lake, databases, and data warehouses. It helps data scientists find relevant training data. The benefits are twofold. First, data catalogs reduce the time data scientists need to identify training data. Second, the AI models get potentially better because the data scientists could use a wider variety of available, relevant training data. We take a closer look at data catalogs in a later chapter.

Data catalogs help to find existing training data. Sometimes, data scientists work on AI solutions for which no useful training data exists. Suppose an AI-driven application should decide based on images whether a steering wheel on an assembly belt can be packaged and shipped to the customer. Training such a model requires having steering wheel images labeled either as "good quality" or "insufficient quality." Collecting and labeling images are time-consuming, tiring, and expensive because training sets have to be large. How long would you need to take 1000 pictures of steering wheels on an assembly line? Some have to be suitable for shipment; others must not. Regarding the defective steering wheels, you need to take at least 20 images for each quality issue.

The good news for data scientists and AI projects is: they can optimize the labeling task with services such as **AWS SageMaker Ground Truth**. The service innovates labeling in multiple ways:

1. Out-tasking the labeling to AWS-managed workers or third party service providers. This out-tasking frees data scientists from labeling large data sets themselves or having to identify and coordinate internal or external persons helping them. Out-tasking works if the labeling does not require too specialized skills. Also, data privacy constraints and the protection of intellectual property can limit such out-tasking.

2. Managing the workforce: The service compiles batches of texts or pictures to be labeled and assigns them to the various internal or external contributors – no need to ask for the status by mail and phone or to organize project meetings.

3. Increase the labeling quality by letting more than one individual contributor label the same pictures or texts.

4. Reduce human labeling effort: AWS can distinguish between "easy" training data items. It can label itself and challenging training data items that a human contributor has to label.

The next phase is "data preparation." It is the tiring, time-consuming, cumbersome work of getting the initial data set in a helpful form for training an AI model. Data scientists can perform this work more efficiently with AI-specific **integrated development environments** (IDEs) such as the already discussed Jupyter notebooks. They speed up the development process by enabling to execute only selected commands of a long script, by easing writing comments and documentation and collaboration.

Auto machine learning (AutoML) promises to automate data preparation, choosing a training algorithm, setting the hyperparameters, and training an AI model. AutoML understands these tasks as (high-dimensional) optimization problems, which a clever AutoML algorithm solves autonomously without human intervention. You provide a table – AutoML returns you a ready-to-use trained AI model.

It sounds like science fiction, but various offerings on the market prove the opposite. Examples are Google's GCP AutoML Tables, SAP Data Intelligence AutoML, or Microsoft Azure Automated ML. It is not clear yet whether and under which circumstances AutoML outperforms or constructs similar good models like the ones "hand-crafted" by experienced data scientists. Independent of that, AutoML is already a game-changer. It speeds up and simplifies creating AI models without needing large data science teams. The Pareto principle (80/20 rule) applies to data science projects as well. It is often better to have a relatively good model quickly with low effort than spending months for the perfect model.

Finally, AI organizations can optimize the deployment phase as well. This phase consists mainly of generic IT engineering tasks. Thus, general improvements such as CI/CD pipelines bring benefits. The AI models become part of the application code or run on a separate AI run-time server. Fewer frictions equal fewer potential errors. Other engineering teams contact the data scientists less often to help them in debugging or fixing integration issues.

Not all presented optimization options help every AI project. The higher on the AI technology stack, the fewer areas do projects and organizations have to optimize. When companies rely entirely on existing, ready-to-use **AI services**, they neither train machine learning models nor deal with training data. They outsourced everything. Nothing is left that can or should be optimized.

Companies relying on **Customized AI** provide training data but delegate the actual training and model creation to external partners. Organizations following this strategy can optimize their work with data catalogs and improved data labeling processes.

Many data scientists and organizations (still) prefer to train the machine learning models using **AI Frameworks** such as TensorFlow. They benefit from data catalogs, improved data labeling processes, or IDEs. They can also use AutoML to improve and (semi-) automate specific data science tasks.

In general, optimizing an AI team's efficiency has two facets. The most important one is to go for a high layer of the AI technology stack if possible. Then, additional optimizations are possible that are relevant for the chosen AI technology layer (Figure 2-20).

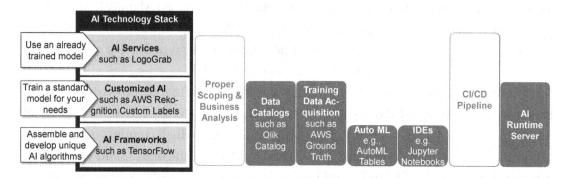

Figure 2-20. *The Big Picture of AI Innovation: Technology Stack, Use Cases, Improvement Options*

Integrating AI Models in IT Solutions

Online shoppers are familiar with the phenomenon. Webshops offer their customers products that perfectly match the contents of the shopping cart: a pair of black socks with a black suit? Click! Pink shoes for the customer with the pink handbag? Click! AI algorithms run in the background. Their aim? Increase sales! To make it work, however, an AI model alone is insufficient. We have to integrate the AI model into a "real world" application such as the webshop.

Data scientists new to big organizations and with a long and extensive academic background tend to underestimate the integration challenge. They start experimenting with a commercial or free AI software package or cloud-based AI services. They want to figure out whether and how to benefit from AI by trying out some use cases. At this moment, they do not (yet) require an interface between the AI component and one or multiple applications. The situation changes once the benefits are evident. The business and the IT management want AI to become an integral part of complex software solutions. For example, an online fashion shop has many features that have nothing to do with AI. The webshop presents fashion; customers can search for items, add them to a shopping cart, and pay. An AI component is simply another feature that "only" determines products for each customer that might be of particular interest to these customers. The AI component might suggest black socks for customers with a suit in their shopping basket and transmit the suggestion to the webshop application. The customer then sees black socks in the browser as a product proposal.

The collaboration between the AI component and the rest of the application is the goal. The starting point for AI – and software solutions with AI functionality – is historical data. Data scientists need them to create AI models, for example, for customer buying behavior. However, it would be a bad idea if data scientists perform their work on a production system. When data scientists query large data sets and run complex, resource-intensive statements, they potentially interfere with daily operations. Thus, decoupling their work by copying or extracting data from the production system and putting the data into a training environment with sufficient computing power is essential.

As soon as the AI model is ready, data scientists or IT engineers deploy the model in production systems. No model is there for eternity. A newer model will replace it after a day or some weeks or months based on more recent data (Figure 2-21).

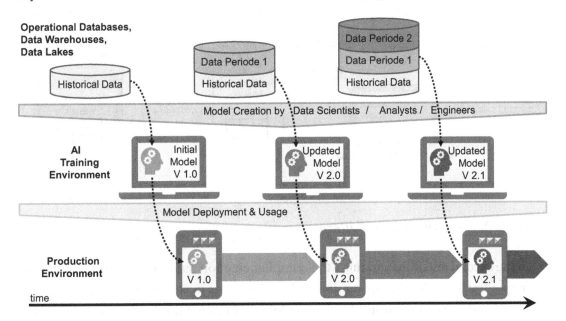

Figure 2-21. *Understanding Data, Model Creation, and Model Usage*

The exact deployment or integration can follow various patterns. The three primary variants are:

- Precalculation

- Reimplementation

- Encapsulated AI component

For **precalculation**, a data scientist develops and creates the AI model, which, for example, determines for each customer the product she most likely buys next. Next, the data scientist or an IT engineer uploads this information to the application. Now, the application knows which items to suggest to the customers when they visit the online shop again later. The application works with the result. It does not know the model. There is no need for integration between any AI component or environment and the application. Model development and evaluation take place independently in the AI training environment, and the evaluation or interference bases always on the most up-to-date model (Figure 2-22).

Precalculation has some limitations if the frequency of model updates and the "real" world changes do not match. For our webshop example, the webshop continues to suggest a neon-green cocktail dress to a customer after she bought one till the next data upload. In such a case, the ads do not generate additional sales – and let the customer feel that the webshop cannot provide a shopping proposal matching her needs.

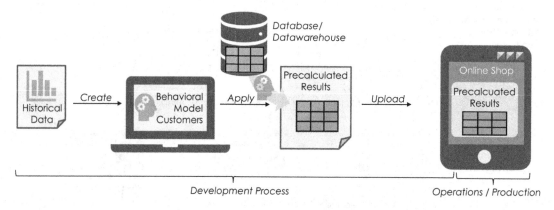

Figure 2-22. *Precalculation Pattern*

In the second and third variants – the reimplementation pattern and the encapsulated AI component variants – the AI component feeds real-time customer shopping data to the model and always gets up-to-date suggestions for online shoppers. Suppose the data analyst creates the model on April 30 in his AI training environment. However, applying the model on user data is deferred to when a customer visits the webshop. Then, her up-to-date shopping history is fed into the model to get product proposals. If she bought a green dress yesterday and extravagant sunglasses an hour ago, the AI component considers this. When a customer visits the webshop on May 17 in our example, the application uses the model created on April 30 and customer data from May 17.

The second and the third variant differ regarding the technical implementation. The **reimplementation pattern** means to program the code again for the software application a second time. Software engineers take the model from the training environment and implement the same model now, for example, in Java. The engineers put the neural network parameters and weights into a configuration file, such that parameter updates do not require a code change. Ideally, a CI/CD pipeline automates the integration. Such a pipeline automatically collects the various sources from all developers and data scientists, creates a deployable software component, and installs it on development, testing, and/or production systems. Thus, when a customer visits a webshop, the webshop contains the neural network, such as any other application feature or code (Figure 2-23).

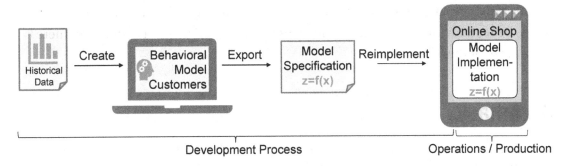

Figure 2-23. *Reimplementation Pattern*

Alternatively, the application landscape might have a dedicated **AI runtime server** in the encapsulated AI component pattern (Figure 2-24). The latter runs the AI models for all the company's applications relying on AI functionality. RStudio Server is such a product. When a customer visits a webshop, the webshop invokes the model on AI runtime server, passing this customer's shopping history. The AI runtime server pushes the shopping history data through the neural network to get a prediction of the customer preferences. Then, the AI runtime server returns this prediction to the webshop, which presents and highlights relevant product(s) for this concrete customer.

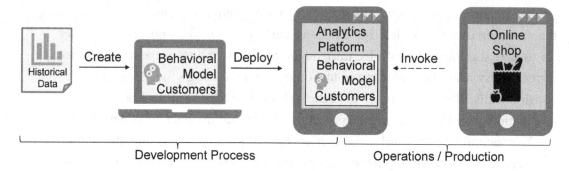

Figure 2-24. *The Encapsulated Analytics Component Pattern*

For this pattern, the AI team has to set up, run, and maintain an AI runtime server to which every data scientist deploys his models.

As a side remark, "models that continuously optimize themselves" sound obvious and visionary at the same time. They are technically feasible. For example, Google's GCP ML Pipeline follows this idea. However, fully self-optimizing and completely autonomous models can have harmful side effects. Suppose AI only suggests extravagant ties with a high margin since the optimization goal is increasing the margin. In that case, the profit might increase in the short term. But at some point, the fashion webshop is perceived as a tie shop, and only customers who want a tie come. Humans notice such trends better than fully automated optimization processes. Even in a world of AI, there is still a need for the human brain.

Summary

This chapter explained every AI organization's potentially most apparent and vital task: delivering an AI project. It elaborated the various facets and layers of AI innovation, scoping AI projects, and how AI models such as statistical or neural network models – the core delivery of every AI project – look in practice. Further, the chapter elaborated on the development process of AI models and the integration in a corporate IT application landscape. With this knowledge, the next big challenge for AI project managers is obvious: How can AI project managers validate whether an AI model behaves as expected and provides the needed prediction or classification quality?

CHAPTER 3

Quality Assurance in and for AI

With AI influencing important decisions in companies, organizations, and our private lives, the quality of AI models becomes a concern for managers and citizens alike. Sales success depends on targeting customers with personalized and fitting products. In the medical area, researchers want to identify Covid-infected persons by letting an AI model analyze how persons cough into a microphone. These two simple examples prove: AI passed the tipping point from harmless, academic experimentation to real-life solutions. The time has come that AI projects require auditable quality assurance and testing. Errors can impact financial results severely or threaten human lives. Data scientists, test managers, and product owners cannot continue treating AI models as magical, always correct black boxes created by brilliant-minded, impeccable, and sacrosanct specialists.

Consequently, this chapter elaborates on how to structure and manage preventive quality assurance by taking a closer look at the following topics:

- **Model Quality Metrics** measuring the quality of AI models and making models comparable. They prove or disprove whether the models are as good and valuable as the data scientists promise.

- **QA Stages** defining a structured process when and in which order projects must calculate quality metrics based on which data. These stages ensure a smooth progression of AI projects towards ready-to-use AI models. They avoid wild, unfocused, and untransparent experimentations.

- **Model Monitoring** continuously checks whether AI models deliver reasonable prediction and classification quality weeks and months after the deployment.

© Klaus Haller 2022
K. Haller, *Managing AI in the Enterprise*, https://doi.org/10.1007/978-1-4842-7824-6_3

- **Data Quality** reflecting the high impact of good data on the prediction and classification quality of the trained AI models.

- **Integration Testing** looking at whether the AI model and the rest of the application landscape interact correctly.

Figure 3-1 summarizes and illustrates the dependencies of these topics. The following sections look at the respective topic boxes in detail.

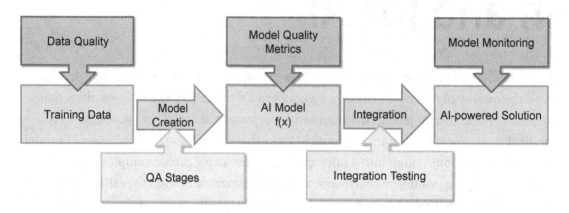

Figure 3-1. *Quality Assurance and Artificial Intelligence – The Big Picture*

AI Model Quality Metrics

A basic understanding of AI quality metrics enables AI project managers to talk "data science" with their team members with postgraduate degrees in math, statistics, or computer science and give them the impression of understanding them. Does the equation $y=67*x_6+54*x_5+23*x_4+54*x_3-9*x_2+69*x_1+42$ provide an adequate price estimation for used cars? Or should we replace 54 with 53 or with -8? Just looking at a linear equation or thousands of weights of a neural network does not allow us to judge whether an AI model works appropriately in reality.

Performance Metrics for Classification

The best known and most widely used quality assurance metrics for classifications – does a customer buy a bag or is this a cat image – is the **confusion matrix**. A confusion matrix allows comparing the quality of models trained with the same algorithm, for example, two logistic regressions basing on differently prepared data. Also, it enables comparing models trained with different algorithms, for example, one with a neural network and one with a logistic regression algorithm.

Figure 3-2 illustrates an AI model for classification that takes images as input and tries to determine whether an image contains a dog. Four of the five images classified as dog images are indeed truly dog images. Thus, the confusion matrix has the value four in the *true positives* cell. The one image wrongly classified as "true"/"dog" results in the value 1 for *false positives* in the confusion matrix. The AI model did classify various images in Figure 3-2 not to be dog images. That is correct for four cases (*false positives*) and wrong twice (*false negatives*).

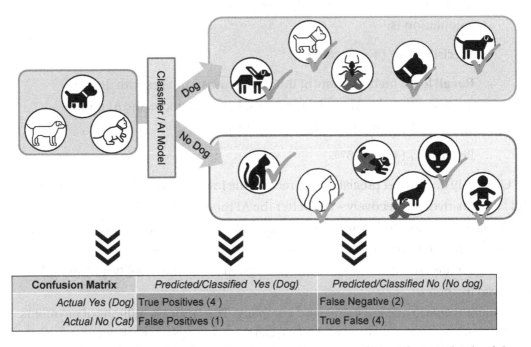

Confusion Matrix	Predicted/Classified Yes (Dog)	Predicted/Classified No (No dog)
Actual Yes (Dog)	True Positives (4)	False Negative (2)
Actual No (Cat)	False Positives (1)	True False (4)

Figure 3-2. *Image Classification – Look at an Image and Decide to Which of the Two Classes It Belongs, Dog or Cat*

The confusion matrix is the foundation for three advanced metrics:

- **Accuracy** is the percentage of correctly classified images. Basically, this is the number of images with dogs correctly identified as such, plus the number of images correctly classified not to contain a dog divided by the overall number of images. In the example, the calculation goes as follows:

  Accuracy = $(4 + 4) / (4 + 4 + 1 + 2) = 8 / 11 \approx 73\%$

- **Precision** reflects how many of the images classified (correctly or incorrectly) as dogs are really dogs. In our dog example, the calculation is:

  Precision = $4 / (4 + 1) = 80\%$

- **Recall** focus on how many of the "true" cases – images with dogs in our example – the AI model identifies. In our example, the AI model identifies four out of six images with dogs:

  Recall = $4 / (4 + 2) = 67\%$

Obviously, the higher precision and recall – the lower the number of false negatives and false positives, respectively – the better the AI model. Optimizing both, precision and recall, is sensible, desirable, and necessary. However, ultimately, projects have to decide which of the two is more important. The choice depends on the usage scenario. It is often a trade-off once the model reaches a good prediction quality and should be improved further.

Sample scenario one is a credit card company. They process millions of payments per day. Their earnings are 0.1% of the transaction volume. So, a CHF 2000 payment generates CHF 2 revenue. A fraudulent payment means that the credit card company has to write off the total amount. Thus, letting one fraudulent payment slip through has a massive impact on revenues. In such a scenario, the credit card company prefers a high recall rate and accepts a lower precision rate. They block as many potentially fraudulent payments as they can identify. Obviously, there are limitations. Blocking too many payments reduces the income, and customers change to competitors if they cannot use their cards adequately in their daily lives.

Scenario two – determining stocks that potentially outperform competitors – is an example of optimizing precision. A hedge fund might prefer an AI model that identifies five stocks outperforming the market over a model that proposes twenty stocks if share prices plunge next week for half of the twenty.

These two sample scenarios illustrate that project managers need a vision of how they position the project in the precision vs. recall continuum. They can even provide a formula for the data scientists in an ideal world, for example, five false positives are equally harmful as two false negatives. Data scientists feed this information as input to the AI model training algorithms.

Classification and Scoring

The discussion about classification model performance gets another nuance when understanding a classification not as a true/false decision but as a **scoring** and **ordering** task. A logistic regression typically returns a value between 0 and 1 for each image. A value over 0.5 means a dog; a lower value indicates that it is not a dog. However, we can also work with the original data between zero and one and interpret the values as an order for their likelihood to contain dogs. To prevent misunderstandings: the values are not percentages or probabilities, just scores!

Figure 3-3 does exactly such an ordering of the images known from Figure 3-2.

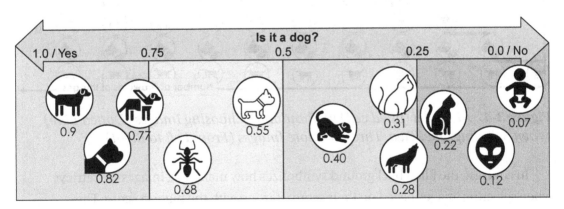

Figure 3-3. *Image Classification as a Continuous Classification Problem Rather Than a Dog/No Dog Phenomenon*

An excellent way to understand this ordering is to think about bubbles in champagne. If you fill a champagne flute and look at the champagne, the bubbles make it one by one to the top – without adding new bubbles. How quickly an AI model moves relevant elements to the front is also a performance metric.

Figure 3-4 illustrates the process. The set of images consist of 6 dog images and 5 other images. When randomly choosing an image, we have a roughly 50% chance to select a dog image. If we choose two images, we can expect to retrieve one dog image; if we choose four, we expect two. The dotted line shows how many dog images we can expect when randomly selecting images.

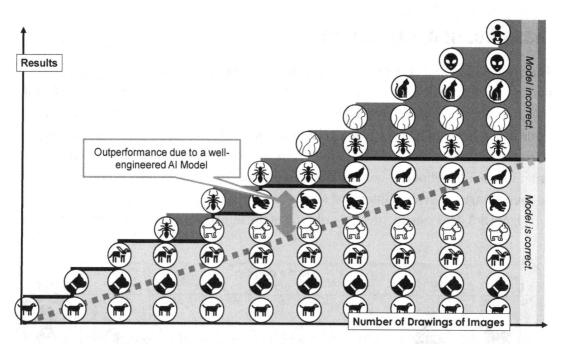

Figure 3-4. *AI Model (filled cell) vs. Randomly Choosing Images (Dotted Line) When Selecting One, Two, Three, or More Images (From Left to Right)*

In contrast, the filled background symbolizes how many dog images we retrieve when selecting images based on their score, starting with the highest score. The indication for the performance of an AI model is how many "dog" images are over the dotted line, which represents the expected performance for randomly selecting images.

In Figure 3-3, the first three images are all dogs – quite an impressive result for the AI model. The probability of getting such a good result when randomly choosing images is only 12%.

Additional Performance Metrics

True positives, false positives, accuracy, precision, recall – these terms and concepts work well for binary classifications. Does the picture contain a dog? Does the client buy a new white suit if we send him a discount coupon by email? We need more varied metrics for other types of insights and AI models with more complex outputs. In the following, we look at metrics for three more scenarios:

- Linear Regression

- Multiclass Classification

- Object Detection, that is, which objects are where on a picture

Suppose you are an AI project manager and run a status meeting. Suddenly, data scientists talk about **R^2**, also named *R squared* or *coefficient of determination.* A good AI project manager has to figure out quickly how to react. Is it time to cheerfully congratulate his team for their breakthrough – or do they need emotional support after a bitter defeat against the world of math? R^2 is a metric reflecting how well a linear regression model reflects reality or the data items in a training data set for prediction problems. So, whereas for classifications, the output is "dog" or "cat," the result for predictions is "We expect 5.7 incoming calls per minute in our call center in the next hour." and not 20.3 and not just 1.

The calculation of R^2 is simple, but the underlying rationale requires some thinking. The first step is calculating the difference between the predicted and mean values and squaring the result for all data points. After summing up the squares, the result is called the sum of residual squares (SQR).

Second, you calculate the difference between the actual value and the mean value and square the results. Again, you sum up the squares over all data points. The result is the sum of squares total (SQT). Then, $R^2 = 1 - (SQR/SQT)$.

What AI project managers should understand is that R^2 is between 0 and 1. The closer to 1, the better the model. High values are possible for technical and physical processes with clear and known variables; for predicting single persons' individual (!) behavior, even 0.1 can be good. However, when looking at larger groups of humans (not at the level of a specific individual), values between 0.4 and 0.8 or even more can be realistic.

An AI project manager's (or AI translator's) task is to put such metrics and numbers into a business context. What advice could the data scientists give to the business? What should the business, for example, do, and how confident are the data scientists that this is the right decision? Increasing sales by 5% might be more than sales managers hoped and dreamed of achieving. On the opposite, good predictions and models might be useless. Do you know the famous forecast from the stand-up comedian George Carlin: "Weather forecast for tonight: dark. Continued dark overnight, with widely scattered light by morning." Sophisticated AI models predicting the obvious are useless. If companies invest in complicated AI models, the models must be better than a naïve and obvious guess.

As always, there are more prediction performance metrics. Worth mentioning is especially the **Adjusted R^2 Metric**. It also considers the complexity of the prediction function. The metric punishes, for example, if the prediction function is not just linear ($y=ax+b$), but also contains quadratic ($y=ax^2+bx+c$) or is even a higher-level polynomial. It punishes incorporating many input attributes such as $y=ax_0+bx_1+cx_2+dx_3+ex_4+f$ instead of selecting the most relevant ones only, such as $y=ax_0+bx_3+c$. The rationale is that complex functions improve the prediction quality but potentially cause overfitting (the latter is an issue this chapter addresses later).

Multiclass classification means that the classification result is not just a yes/no, but can be a dog, cat, turtle, or giraffe. In a business setting, such an AI model might predict the next best product for a customer. Should the bank advisor try to sell the client an equity fund, a mortgage, or a golden credit card?

Multiclass classifications need an expanded confusion matrix, as Table 3-1 illustrates. Each class requires a matching column and row. Again, the matrix allows comparing predictions and reality.

Table 3-1. Confusion matrix example for multiclass classification

	Predicted: Credit Card	Predicted: Mortgage	Predicted: Equity Fund	Predicted: Savings Plan
Actual Bought: Credit Card	189	5	8	7
Actual Bought: Mortgage	12	25	5	4
Actual Bought: Equity Fund	12	13	58	24
Actual Bought: Savings Plan	7	1	51	125

For multiclass classifications, the metrics definitions change slightly – and more metrics variants exist. Here, we explain two: (multiclass) accuracy and balanced accuracy. Comparing both gives project managers an idea of what different purposes or foci metrics can have.

Accuracy for multiclass classifications is the portion or percentage of correctly classified or predicted items. In the example, we divide the sum of the items in the diagonal by the sum over all cells:

$$Acc_{multiclass} = \frac{189 + 25 + 58 + 125}{189 + 5 + 8 + 7 + 12 + 25 + 5 + 4 + 12 + 13 + 58 + 24 + 7 + 1 + 51 + 125}$$

The predictions of the frequent classes dominate the result. Suppose some classes appear more often than others. In that case, the classification quality for the scare classes has a low impact on this metric. For example, different cancer forms are illnesses not many have at a defined point in time. An algorithm that predicts "no cancer" for each and every patient performs close to 100%. The very few cancer patients have (nearly) no impact on multiclass accuracy. Multiclass accuracy is worthless for this scenario. A metric that deals better with scare but important classes is the **balanced multiclass accuracy** metric. It averages the metrics of each class:

$$Acc_{balanced} = \frac{\frac{189}{189+5+8+7} + \frac{25}{12+25+5+4} + \frac{58}{12+13+58+24} + \frac{125}{7+1+51+125}}{4}$$

In other words: All classes have the same weight in case of this metric, no matter how frequent any of them is. The performance on multiple infrequent classes outweighs issues on one or two dominant classes for this metric.

Multiclass and balanced multiclass accuracy are two examples of quality metrics for multiclass classifications. None of them is per se better than the other. The driving factor is the business question the AI model wants to solve. AI project managers must understand which metrics the data scientists use and why. It is essential for delivering business value. Choosing inadequate metrics means that the created AI model might not deliver the needed business value. However, it is not always easy to make the right choice, especially for a complex business question. Still, it is a good idea to start with established metrics. There is a reason why they are widely used: they are often helpful.

Measuring the performance of **object detection** goes much further than simple multiclass classification. First, images can contain more than one object, for example, a glass, a bottle, and a chair (see Figure 3-5). Second, the AI model also determines bounding boxes for each object. The boxes mark the location of the identified objects on the image. Typically, an object counts as correctly detected if bounding boxes overlap at least 50%.

In Figure 3-5, the AI model identifies the bottle, the glass not. While the object class "glass" is correct, the overlap of the shapes is too low. Finally, the chair is not detected, and there is no car where the algorithm detected one. So, how good is the AI model? One metric is defined as dividing one by four, resulting in a value of 25% (Intersection over Union).

To conclude, the examples for performance metrics for predictions, multiclass classification, and object detection illustrate the variety of existing metrics for the different AI challenges. Some are more complicated than others, but AI project managers invest their time wisely when they insist on understanding what their data scientists optimize. The AI project success depends on whether the metrics the AI model optimizes match the relevant business performance metrics.

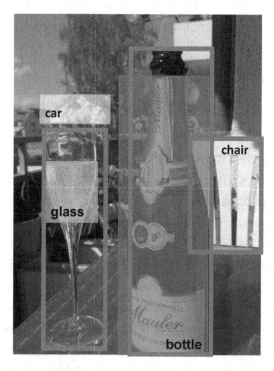

Figure 3-5. *Object Detection Illustrative Example with Various Objects. Green Bounding Boxes Are the "Correct" Ones, Whereas the Transparent Boxes Are the Limits the Model Identified*

QA Stages in AI Model Engineering

When data scientists present their AI project model and the precision and recall or R^2 values, should he run to his sponsors and tell them they made it? Is the model ready for production deployment?

The answer is a clear "no." A single metric is never enough. The project manager needs at least two to three metrics and confidence that the training data was used adequately to produce the metrics. Figure 3-6 outlines a process of how to structure the quality assurance for AI models. A good model is not the lucky result of random parameter or weight changes if the metrics are not as good as needed or expected. Getting a good model means starting with multiple candidate models. The project drives them through the various QA stages. Eventually, the chosen one makes it through all stages because it is the first good model getting there or a candidate model superior to all others.

Figure 3-6. *Understanding Quality Assurance in the Process of Turning Data into AI Models*

Perfect but Worthless Model Metrics

Projects should avoid worthless metrics, that is, metrics that do not help to understand the model performance. Some car manufacturers impressively perfected the concept of useless metrics, which turned out to be a bad idea. They optimized car engines a little bit too much. All cars met the regulatory emission benchmarks because the car adjusted the engine when noticing to be in an emission test. Emissions under normal traffic conditions were completely different and much less eco-friendly. It was not for the good of the involved managers and engineers. AI projects should avoid – even unintentionally – producing too-good-to-be-true metrics as the car manufacturers. Such issues happen when mixing up the data for training and for assessing an AI model.

To add a second nuance: Do you know the fastest way to implement a 100% perfect model if using the same data for training and quality assurance? Program a component that checks whether the model input data corresponds to an entry in the training set. If so, take the correct answer from the training set. Otherwise, return a random number. Your model achieves 100% for precision and recall on the training set – and is entirely useless. The only way to avoid running unintentionally into such issues is to split your training data into three data sets: training, validation, and test data sets – and to follow a strict quality assurance procedure this chapter explains.

The Training, Validation, and Test Data Split

The purpose of a **training dataset** is to build and train a machine learning model with this specific data. There are many ways to optimize and fine-tune AI models, such as trying out various mathematical function(s) or changing the number of hidden layers for neural networks. Data scientists perform these optimizations using the training data (only). In contrast, the purpose of the **validation dataset** is to compare multiple candidate models and to check for overfitting. The final test before actually deploying a model is a final assessment using the **test dataset**.

The split of the dataset depends on the context. Datasets often consist of several hundred or a few thousand data points for predictive analytics and statistics scenarios. The typical **split ratio for training/validation/test datasets** for them is 60%/20%/20%. In big data, machine learning, and neural networks, there are often millions of data points. Then, a 98%/1%/1% split ratio makes sense. Validation and test datasets of around ten thousand data points are sufficient. Dividing the existing historical training data into these three subsets is essential before starting with the model training.

Assessing the AI Model with the Training Dataset

The first check of a newly trained AI model is how it performs on the training set. If AI models fail directly here, they are unlikely to perform better with live data in production in the "real world." So, this is the first real QA step for a new AI model.

Assessing the training model with training data means feeding the training data set into the model and check whether the model's predictions are correct. Does the model predict the bank customers to be interested in a credit card that got the product? The data scientists calculate the precision, recall, and accuracy metrics for classifications (or R^2 for predictions) and verify whether the value is acceptable for the business. If the company needs a 95% recall to make a business case a success and the model reaches only 85% on the training set, moving forward with this model is a waste of time and money. The AI project team has to fix and improve the model first.

The visualizations in Figure 3-7 illustrates potential root causes for failing models. The curve represents the AI model, which divides the area into two subareas or classes. All cat images are in the upper left and middle part, all dog images in the lower area – 100% precision and recall. In Figure 3-7 (a), the curve separates the data set(s) entirely correctly. It is the perfect case, but usually, the situation sketched in Figure 3-7 (b) is closer to reality. The illustration helps understanding two reasons why AI models might not even perform well on their own training data.

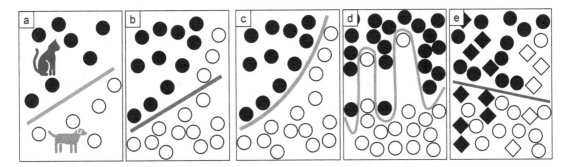

Figure 3-7. *Scenarios for Classifications: How Good Does the Curve Representing the Classification Function Separate Cat (Black Circle) from Dog Images (White Circle)?*

The first potential root cause is **missing crucial information** and attributes. A picture of a car in a parking lot is not sufficient to determine whether the vehicle has enough gas to drive thirty kilometers. Thus, even if a distinction of two classes is possible in theory, models fail for specific data if they do not have all information. The solution is obvious: add attributes to the data points of the training data and retrain the model.

The same figure also represents a second reason for an AI model failing on training data: The underlying **model** function is too **simplistic** (**"underfitted model"**). Linear regression does not work well for complex, non-linear challenges. In our illustration, the AI model draws (only) one straight line to separate classes. Only a curve can separate the areas correctly (Figure 3-7, c). If the mathematical function does not match the problem, the AI model performs poorly. Data scientists fix such issues by moving from linear regression to a neural network or adjusting or extending the neural network and then retrain the model.

To conclude: quality issues causing insufficient AI model quality on training data can have two root causes, underfitting and missing information. Data scientists have to dig deeper and potentially experiment to understand the details. But once models work well on the training data, it is time for the next QA stage, which takes advantage of the validation data sets.

Assessing the AI Model with the Validation Dataset

When the AI model performs well on training data, the next step is testing it with the validation dataset. We do precisely the same as in the previous assessment, just with different data. This time, we feed the validation data set. The validation data is data *not* used for training the model. This measurement aims to detect **overfitting** or a **high bias** in the training data. The symptoms are the same – good performance (e.g., recall, precision, R^2) on the training set, poor performance on the validation set.

In the case of **overfitting**, the model does not generalize well. The AI model reflects all the oddities of the sample data but does not generalize, which is needed to perform well on unknown data. For those with a strong math background: If you have a training data set with 1001 data points, you can fit a 1000th-degree polynomial as a prediction function. The function works perfectly for the training data but is completely erratic everywhere else. Figure 3-7 (d) visualizes this phenomenon. The curve separates the dog and the cat images well for the training data. However, the border between the areas is so erratic that the model classifies images similar and close to training images differently.

Figure 3-8 provides another perspective on the underfitting/overfitting challenge. On the left side of the figure, the model is underfitted and too simplistic. The model error on the training and the validation data set is high. When the model gets more complex, its performance increases, that is, the error drops for the training and the validation data. When the model gets more complex, the model error gets lower and lower for the training data, whereas the model error on the validation data rises. We are overfitting the model. The challenge for data scientists is getting close to optimal model complexity, the tipping point between under- and overfitting. Regularization is one option to ease their work. It means penalizing more complex models when comparing model quality.

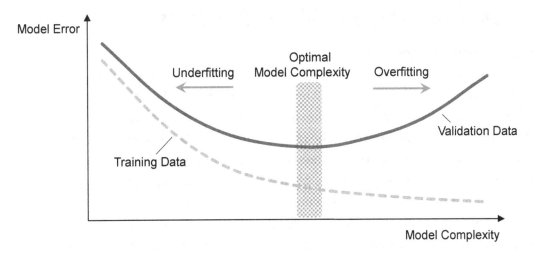

Figure 3-8. *The Model Complexity vs. Model Error Dilemma*

Biased data is the second potential root cause for good model performance on training, but not on validation data. Biased data means that training and validation data (and potentially the reality) have different data distributions and characteristics. In our cat-dog example, one data set might consist of a well-crafted set of dog and cat pictures. A second and a third data set might result from crawling the Internet and Wikipedia for images related to cats and dogs. Obviously, training a model with one of the data sets results in a good performance on this specific data set, but not necessarily when testing with any of the other sets. Without a budget for manually reviewing images, automatically collected training data probably contains images labeled as a cat that contain cat food or litter boxes for pets. All three approaches are legitimate to get data for training and validation and test data sets. The bias issue emerges when not mixing the images from the different data sets, but one becomes the training, the next the validation, and the last the test data set. The characteristics of the image sets most likely differ too much.

Figure 3-7 (e) illustrates the bias issue. A vertical line separates the two classes quite well when training a model with the circles. However, suppose the distribution of the validation data – symbolized with the rhombs – differs. In that case, the model does not classify them correctly.

Hopefully, your AI project trained one or more candidate AI models which perform well on the validation test data set. Then, the AI model is ready for a final check with the test data before production deployment.

Assessing the AI Model with the Test Dataset

The final assessment is with the test dataset. An AI model in this stage passed the assessments successfully with the training and validation data. Still, it can fail on the test data with poor metrics values. Then, the model is probably overfitted to the validation data under the assumption that validation and test data sets are similar.

This explanation might surprise. The purpose of the validation stage is to detect overfitting, yet we face this issue in the test stage. The reason is simple: The validation stage detects overfitting to training data, not to validation data. Since the model is now in the test state, it performed well (also compared to other candidates) with validation data. Obviously, an overfitted model has a good chance to perform well and to get to the test data stage. How to fix this shortcoming? Increase the validation data set to make overfitting less likely, repeat the validation test, and choose a different model for the final assessment with test data.

Monitoring AI Models in Production

With the deployment of an AI model to production, research-oriented textbooks stop. What could you expect more than a high-performing model? In practice, the business expects a well-performing model not only for a day but for weeks and months. If the quality drops, retraining becomes necessary. At deployment time, the first question might be whether the model provides the classification and prediction quality required for the expected business benefit. Furthermore, the business might pose a second question: How can we ensure that the model offers improvements and savings over the long term. Even if they do not ask right now, they will clearly communicate their expectations if business benefits vanish and sales figures plump due to a not-so-good-anymore AI model.

A forward-looking AI project manager addresses such uncertainties and obvious necessity by implementing monitoring functionality as part of the project. It is a paradigm shift compared to the traditional QA approach in IT departments. The traditional QA strategy tests rigorously before going live. Afterward, the rule is: never touch a running system. AI models need a different approach. They might degenerate and become outdated within hours, days, weeks, or months. It is often a gradual process, but there are clear, measurable indicators:

- A significant change in the input values, for example, changes in the mean value or the distribution of input variables. Do shopping baskets contain 80% swimwear instead of 80% products of the winter clothing category?

- A significant change in the output values of the statistical model. Does the distribution of the classification output change, for example, from 30% dog pictures to 45%? Will only 20 products be proposed to all customers over one day instead of 1000 different products per day previously?

- Declining forecast quality: Are 20% fewer customers adding the product proposed by the AI component to their shopping cart than a month ago?

Of course, an application must collect these key figures so that business users or data analysts can recognize changes or an automatic alarm is possible.

An (more complex) alternative to monitoring indicators is retraining the actual model from time to time or even every night. A new model is generated with the then-current data and compared with the previous model. In the event of significant deviations, data scientists replace the older model with the new one. How often do you generate a new model for control purposes – and whether you can automate this model creation – is ultimately a commercial question.

Data Quality

The problem with cheating in exams by copying your neighbor's answers is simple. Typically, the least-prepared students do not sit next to the best and brightest. Their neighbors might also not have a clue – and replicating wrong answers and building on them usually does not end well.

Data scientists face similar situations. They create AI models, sometimes too naïvely, assuming that the training data is in good shape. A recent study found publicly available training sets used in academia and research to have issues with 0.15 to 10.1% of the training set elements. These numbers should be a warning for data scientists and AI project managers to take the topics seriously during the CRISP-DM phase "data understanding." Projects benefit from looking at data quality right from the beginning.

The literature is full of good ideas and frameworks. Implementing them and ensuring good data governance practices is often a real challenge. Luckily, this is not the responsibility of data scientists. They are not the ones who should get involved in extensive data cleaning exercises. However, they might have to do some data preparation with limited improvements themselves. From an AI or data science perspective, data scientists and AI projects can restrict their focus on three main data quality aspects (Figure 3-9):

- Technical correctness

- Match with reality

- The reputation of the data

Figure 3-9. *The Three Dimensions of Data Quality*

Technical Correctness

Data is correct from a technical perspective if it matches the defined data model. Mandatory fields have values, expected relationships and links between data sets exist, and data types and actual data match. IT specialists from the database world sometimes forget one fact: not all training data for AI models intelligence comes from tables with strictly enforced schema constraints such as NOT NULL or UNIQUE. Log data, JSON files, or manually filled Excel sheets often are in a less strict format. Plus, some documents are broken for whatever reason, or attributes are missing or filled with placeholders such as "xxxxxx" if not known when creating the row. Humans get pretty creative when forms do not match their needs and reality.

Just skipping – instead of improving – technical inaccurate training data can have unexpected severe side effects. Suppose you create a model for customer behavior in the EU. The various source systems together deliver 500,000 data items for your training set. 35,000 miss multiple attributes. The data scientists suggest skipping them. What should you say or do as an AI project manager under pressure?

Skipping 35,000 from 500,000 data points is not a significant quantitative loss. Still, the impact on the model quality can be severe. Suppose the 35,000 are not equally distributed over the data set, but are the complete German market data. If training a model without such German data, the model probably works nicely on the training data. The metrics might look perfect even for the validation and test data when this data also comes from data that skipped the incomplete rows. However, once in production, the model probably performs poorly for German customers. Thus, addressing technical data quality by removing data requires a second thought and at least a rudimentary analysis before removing the rows from your data set. Otherwise, you might skip essential information.

Data Matches Reality?

About 100 million AUD$ just for extra legroom on a flight was what the Qantas app showed a passenger as the price. In the end, he was charged 70 AUD$ on his credit card. It is an extreme example of discrepancies between data in two systems. Similar, though usually less severe, differences can exist when comparing data in a training set and the reality.

AI project managers and teams cannot validate each and every data item. Still, they can verify various aspects by checking individual data items or looking for hints for quality issues. They could, for example, look at the following:

- Is the provided data, such as street names or sales figures, correct?

- Is the data complete and free of redundancies, for example, there is precisely one booking for each customer purchase – and not two or none?

- Is the data consistent? If, for example, two attributes store the same information, such as the tax location, are they always the same?

- Does the distribution of the training data reflect the reality – or do we have in our training data much more pictures of expensive cars than of the various standard entry-level cars?

- How quickly does a change in the real world result in an adaptation of the data? Does it show up in the training set one hour later, or only next month if customers make purchases?

Measuring such objective criteria helps understand when to rely on specific data sources and to detect if sources have discrepancies and (timely/delay-caused) inconsistencies.

Reputation of Data

The third dimension differs from the first two. It puts human judgment in the foreground. What do recipients of the data (or the recipients of your AI model) think about the data quality of the training data sources? They might not have a short-term alternative if they question your data sources and, thus, the models you create. Eventually, however, they find other options – other teams, doing something "on top" with Excel, or getting a service from an external provider. I learned this lesson on my very first client engagement. We had a consulting mandate within a large IT service provider. I was trying to draw a diagram with all the systems storing and processing data. One system responsible proudly told me that his data is of top quality and always up to date. Well, one of the recipient teams implemented another solution to clean the data before using it.

The system responsible was neither a liar nor negligent. With different needs and usages, the same data can be anything from "not useful" to "excellent." An attribute "sold items" can count the number of items which customers put in their shopping basket – or the numbers of shipped items and not returned. If you are in the controlling department, your focus might be more on correct numbers than on valid shipping addresses. Data quality has a recipient-specific subjective component. This aspect is vital for AI projects. It can impact the acceptance of the AI model if you use known-to-be-good or known-to-be-often-wrong sources for your training data.

QA for AI-Driven Solutions

AI models can support one-time strategic (management) decisions – or optimize and improve repetitive, operational processes. In the latter case, a good model alone is not enough. It must work as part of an AI-driven software solution. Thus, quality assurance must cover three areas:

1. Testing the software solution without the AI component. Is the GUI working? Are calculations correct?

2. Quality assurance of the AI model. Do predictions and classification models work?

3. Verification of the integration of the AI model into the software solution.

The first point relates to standard software testing. Are all features correctly implemented according to the specification? Do the shopping cart feature and the payment process work in the fashion shop app? Is the app not crashing? The IT literature covers testing broadly. Companies have standards and guidelines and established test organizations with experienced testers and test managers. Thus, the first point can be assumed to be solved nearly everywhere, whereas the previous pages covered the second point, QA for the AI model. This last section focuses on the third aspect: potential issues specific to integrating and coupling an AI model into a complete software solution. We have the fashion app, and we have an AI model recommending items to customers – do they work properly together?

Quality assurance and testing measures for the integration depend on the chosen integration pattern. The previous chapter presented three main patterns. The first and most simple is **precalculation**. This pattern does not require any technical integration between the AI component and the software solution. Instead, data scientists or IT operations specialists feed up-to-date data into the AI model, for example, the shopping history of the fashion app customers. Then, the AI model predicts for each customer individually the fashion items she is most likely to buy. These results are written in a CSV or Excel file. A data scientist or IT operations specialist uploads the file, for example, to the CRM tool. The CRM tool sends out personalized push messages to the customers. The messages aim to make the customers curious to look (and buy) the suggested fashion items.

The integration itself does not require any testing, just some validation of whether the generated files match the format expectation of the CRM tool. Since this is a manual process, checklists or a four-eyes principle reduce the operational risks, especially for repetitive processes.

The second integration pattern is **model (re)implementation**. Data scientists provide the model, for example, in Python or R. Then, developers of the actual software solution take over. They add the code as it is or recode the same logic in Java or C#. The model becomes part of the software solution's source code. This pattern requires dedicated test cases to address two concerns.

The first concern relates to *mistakes in the reimplementation*. Recoding a code fragment in a different language poses the risk of mixing up parameters when invoking a function. Also, overlooking and forgetting to reimplement single lines of code is a risk, so is mixing variable names. Test cases with sample input/output data pairs address this risk. An example is to have a test case based on a specific dog image. Suppose the output classification value was 0.857 in the AI model training environment. In that case, a reimplemented AI model in the software solution must produce precisely the same result.

Also, checksums on configuration, metadata, and weight parameters help when replicating neural networks. The sum of all parameter values – a semantically irrelevant number – must be the same. More advanced metrics detect swapped parameter values or switched rows.

The second concern relates to the actual connection or *interface*, especially regarding the parameter usage. Is the first parameter the last bought product ID or the quantity – or is it the other way around? Does "1" mean that the image contains a dog – or does "0" represent the dog information? Simple mistakes result in a useless software solution and ruin, for example, customer experience and sales numbers of the fashion shop app. An AI project manager should double-check if the testers of the software solution also address these quality risks.

The third integration option is an **AI runtime server** such as RStudio Server. The idea is to provide one server on which all AI models in the company run. When a software solution wants to integrate an AI model, it invokes the specific model on the AI runtime server. Potential testing needs are, again, whether the parameters are used correctly for service invocations. Then, the model management becomes more relevant. Is the correct model invoked or deployed, that is, dog detection, and not the model for cats? Is the most recent winter fashion items catalog the base for the model in use – or the beachwear fashion catalog from two years ago? When implementing this integration pattern, software developers and data scientists can work entirely independently and (nearly) without communicating with each other. Thus, checking whether the software solution invokes the correct model with the proper parameters becomes even more crucial in such settings.

Summary

AI offers many organizations new optimization opportunities. At the same time, there are new challenges for testing and quality assurance. Confusion matrixes or the R^2 value – metrics are the base for determining an AI model's quality. They are at the core of any proof-of-readiness assessment. Measuring an AI model's performance against training, validation, and test data provides a structure for the model training and enables AI project managers to track the project's progress.

In contrast to traditional software with a "never touch a running system" attitude, AI models require monitoring. Are they still adequate some weeks or months later, or do data scientists have to train a new and improved version? The key for any convincing AI model is, however, correct training data. AI projects have to take not only the data quantity but also its quality seriously.

For sure, there are also additional non-functional-requirements-like necessities such as regulatory needs, ethics, or explainability. We cover these quality aspects in the next chapter.

CHAPTER 4

Ethics, Regulations, and Explainability

The laissez-faire years for big tech companies, innovative start-ups, and companies exploiting the benefits of IT innovations are coming to an end – also when it comes to AI. Policymakers discuss regulations already for years. But now, they take action. AI organizations and companies have to ask themselves – for ethical reasons – whether they should create the AI models they work on and whether their projects are even legal. They can no longer focus just on developing superior AI models, verifying their quality – and ignore everything else.

While AI ethics sounds like an ivory tower topic, it is not. First, most companies fear negative press and public outrage. They avoid reputational risks, especially when some experts play around with technology without clear business needs or apparent benefits. Second, AI regulations are in flux. Today's AI ethics discussions are the source of tomorrow's AI regulatory necessities. These discussions allow AI organizations to forecast what regulators might demand next. It does not need a fortune-teller to foresee that AI regulation will be a dominant topic in the 2020s.

Consequently, this chapter discusses regulatory and ethical topics. A third topic – explainability – complements these two topics. It is a concept or an AI model property helping to meet specific AI ethics and regulatory demands.

AI Ethics

Suppose you want to buy a new apartment. You ask your bank for a mortgage. They decline. The banker tells you their AI calculated a 75% probability that you get divorced within the next two years. So, no mortgage for you. Would you consider this prediction as ethically adequate? In more than a decade working with Swiss banks, I learned

© Klaus Haller 2022
K. Haller, *Managing AI in the Enterprise*, https://doi.org/10.1007/978-1-4842-7824-6_4

that the big pink elephant in the room is the divorce risk. No banker talks about the topic – especially not with customers – though rumors are clear: divorces are the most crucial reason for mortgage problems. It is a big dilemma for banks. Like most other companies, organizations, and corporations, banks aim to be seen as reasonable and responsible, not as the source of family disputes or escalating relationship problems. Most companies have higher ethical standards than "Don't be evil." These pledges have implications for their AI organizations. They have to be aware of three potential areas where ethical questions might interfere with their daily work and long-term business or IT architectural ambitions.

The Three Areas of Ethical Risks

Ethical risks for AI fall into three categories (Figure 4-1): unethical use cases, unethical engineering practices, and unethical AI models.

Unethical use cases such as calculating divorce probabilities are not AI topics. They are business strategy decisions. Data scientists can and should require a clear judgment on whether a use case is acceptable or not. This judgment is with the business, not with the IT or AI side of an organization.

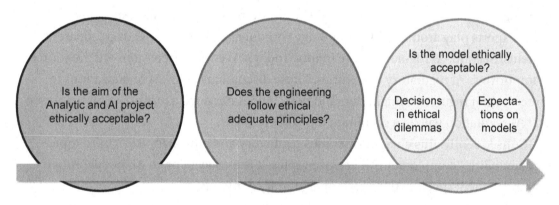

Figure 4-1. *A Lifecycle Perspective on Ethical Questions in AI*

Ethics in engineering relates to the development and maintenance process. For example, Apple contractors listened to and transcribed Siri users' conversations enabling Apple to improve Siri's AI capabilities. This engineering and product improvement approach resulted in a public outcry since society sees the approach as an invasion of the users' privacy.

Finally, the **ethical** questions related to the **AI model** and its implications for customers, society, and humans are the true core of any AI ethics discourse. When algorithms decide who gets the first Covid vaccination and governments outsource the AI model development to a controversial IT service provider, AI decides about life and death – and AI ethics is no longer an ivory tower topic.

Handling Ethical Dilemmas

Ethical dilemmas are on a different dimension than inconvenient management decisions. Should a social media web page post statements of a well-known agitator to attract clicks? Should a start-up harvest public domain images on a large scale to save money instead of making their own photos? These are inconvenient questions for managers, who have to balance ethical behavior and standards vs. profit, loss, and risk. However, real ethical dilemmas are more severe. They relate more to the AI models themselves. Training an AI model means measuring how good a system is. It means quantifying results and optimizing or minimizing loss functions. These loss functions are at the core of **ethical dilemmas**, challenging situations for which no undisputable choices exist. The classic example is the **trolley problem**. Ahead of a trolley are three persons on the track. They cannot move, and the trolley cannot stop. There is only one way to stop the trolley from killing these three persons: diverting the trolley. As a result, the trolley would not kill the persons on the track. It would kill another person that would not have been harmed otherwise. What is the most ethical choice? If the "damage" of killing someone on the track is "-5", what should be the penalty for killing an innocent person – also "-5" or better "-10"?

The trolley problem is behind many ethical dilemmas, including autonomous cars deciding whether to hit pedestrians or crash into another vehicle. When AI directly impacts people's lives, it is mandatory to understand how data scientists train the AI models to choose an action in such challenging situations.

Data scientists can generally design an AI component dealing with such situations in two ways: top-down and bottom-up. In a bottom-up approach, the AI system learns by observation. Since ethical dilemmas of this type are pretty rare, it can take years or decades until a driver faces such a dilemma – if at all. Luckily, most of us never had to decide whether to overrun a baby or a young boy. Observing the crowd instead of working with selected training drivers is an option to get training data quicker. The disadvantage is also apparent: the crowd also teaches bad habits such as speeding or cursing.

Alternatively, data scientists can follow a top-down approach with ethical guidelines guiding through critical situations. The best-known example are Asimov's laws for robots:

1. A robot may not injure a human being or allow a human being to come to harm.

2. A robot must obey orders given to it by human beings except where such orders would conflict with the First Law.

3. A robot must protect its own existence as long as such protection does not conflict with the First or Second Law.

Asimov's laws are intriguing but abstract. They require a translation or interpretation for deciding whether a car hits a pedestrian, a cyclist, or a truck when an accident is inevitable.

In practice, however, laws and regulations clearly define acceptable and unacceptable actions for many circumstances. If not, it gets tricky. Is the life of a 5-year-old boy more important or the life of a 40-year-old mother with three teenage daughters? An AI model making a definitive decision on whom to kill is not a driver in a stressful situation with an impulse to save his own life. It is an emotionless, calculating, and rational AI component that decides and kills. For AI systems, society's expectations are higher than for a human driver. What complicates matters are contradicting expectations. In general, society expects autonomous cars to minimize the overall harm to humans. There is just one exception: vehicle owners want to be protected by their own vehicle, independent of their general love for utilitarian ethics when discussing cars in general.

Implementing top-down ethics in AI systems requires close collaboration of data scientists or engineers with ethics specialists. Data scientists build frameworks that capture the world, control devices, or combine ethical judgments with the AI component's original aim. It is up to the ethics specialists to decide how to quantify the severity of harm and make it comparable – or how to choose an option if there are only bad ones. However, the good news for AI specialists is: ethical dilemmas are exceptional cases.

On Ethical AI Models

Most data scientists have to deal with AI ethics when the question emerges whether their AI models are ethically adequate. These models impact people's lives, be it a credit scoring solution or a face detection model for automated passport controls at airport customs. There is a lot of ongoing research, but two concepts stand out for AI organizations in an enterprise context: bias and fairness (Figure 4-2).

Bias was already a topic in the quality assurance chapter. The concept looks at the training data and its distribution. We call training data biased if its distribution deviates from the one in reality. If a society consists of roughly 50% male and 50% female persons, the training data distribution should be similar. If the training data for face recognition consists of 99% male pictures, the created AI model most probably works well on them. In contrast, the model is most likely unreliable for females.

What makes avoiding bias an easy-to-get-support-for measure is that removing bias and raising ethical standards brings a business benefit: the AI model gets better.

Fairness is a concept demanding that the trained AI models behave similarly well for all relevant subgroups, even for small ones. Suppose a company has 95% female employees. Such a gender imbalance impacts the AI model training. The training could optimize the accuracy for female employees and ignore the male ones and still achieve high accuracy. When 95% of the employees are female, and the algorithm identifies 99.9% of the female population correctly, but only 10% of the remaining 5% male employees, overall, the AI model identifies 95.4% correct – not bad, at least if you are female, not male. However, the ethical concept of fairness rejects such a model. It would be unthinkable that every male employee must show a passport at the reception, whereas all female employees can use their badge to get in.

In an enterprise context, fairness and bias have one crucial difference. In contrast to bias, fairness causes costs for the business. Fairness means not using the best AI model but a model that works equally well for all relevant subgroups, for example, by overrepresenting small relevant subgroups in the training set. To illustrate this, we extend the example from above. The best AI model might identify 99.9% of the female employees but only 10% of the male ones. The best fair (!) AI model might have an accuracy of 90% for both male and female employees. Such a "small" change of the overall accuracy can be cost-intensive. Using a 95% accurate automatic passport controls at an airport requires the customs staff to check the passport of one out of 20 travelers. An accuracy of 90% means checking 10% of the travelers, requiring to double the customs staff.

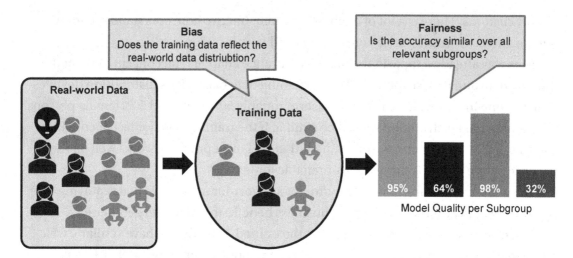

Figure 4-2. *Understanding Bias and Fairness*

AI Ethics Governance

AI ethics is a topic with many stakeholders, from data scientists, AI managers to business specialists needing AI models and corporate social responsibility teams. It is a highly emotional topic, too. Organizations risk bitter fights about what is adequate and what is morally acceptable and who decides what if not working out crystal-clear who or which board makes which decisions concerning the three areas: (un)ethical use cases, (un)ethical engineering practices, and (un)ethical AI models

The Google/Timnit Gebru controversy illustrates an easy-to-oversee pitfall for AI organizations. Gebru's reputation bases on earlier work where she unveiled that IBM's and Microsoft's face detection models work well for white men but are inaccurate for women of color. In other words: She found flaws in the model caused most probably by biased training data. Such flaws are embarrassing, but knowing these shortcomings enables data scientists to build better models.

While with Google, she co-authored a paper that criticized Google's work on a new language model for two reasons. First, Google trained its model with texts available online. Such texts are obviously (partially) biased, discriminating, or even openly racist. When the training set also contains such texts, the model reflects them partly as well. UN's Autocompletion Truth campaign illustrated this in videos. They let Google Search's autocomplete function suggest which words should come next. If "women cannot" results in "women cannot drive," this reflects prejudices in our use of the language. The autocomplete prediction works correctly, but the result is questionable. Gebru was and is right: if you use online texts, you replicate today's injustice and unethical tendencies. The underlying demand of such criticism is: language models should not reflect our language use today, but a better, purified version of ourselves and our language – free from prejudices and discrimination.

This criticism, at the same time, questions the engineering method. Should Google use online texts from the web, which do not incur costs and reflect our language, or should you start filtering and preparing training texts to exclude potentially problematic texts? Suddenly, the criticism impacts the speed of the AI model development and the business case due to the potentially high effort needed to filter and rework texts used as training data.

Her second criticism addresses the massive energy consumption of the language model. This criticism has nothing to do with the ethical properties of AI models, such as bias or fairness. It questions whether Google should develop an AI model that the management sees as essential for Google's future business. Weighting business strategy and its impacts on ecology is a valid ethical question, it is just more on the business side than at the core of AI ethics. Figure 4-3 illustrates this shift in Gebru's research.

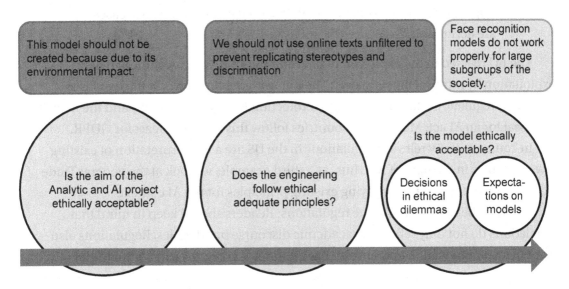

Figure 4-3. *Understanding the Google/Gebru Controversy*

Companies and their AI organizations can learn from this controversy. Which manager or board decides about which (of the three) ethical aspects in the AI sphere? What are the competencies of corporate social responsibility or business ethics teams? What does the top management decide?

AI ethics is essential from a moral perspective, for the corporate culture, and for the organization's reputation – not to speak of the already existing or expected to come regulatory demands. Thus, a frictionless integration in the company and its corporate culture is necessary right from the beginning.

AI and Regulations

Adhering to ethical standards is highly recommended, though optional. Adhering to regulations is a must. And when it comes to regulations, policymakers around the globe passed laws over the last years and are preparing additional ones that impact AI organizations directly. These regulations come in two forms:

- **Data protection laws** defining which data companies can store, how they can use the data, and which additional conditions they have to meet (e.g., transparency, right to be forgotten).

- **AI algorithm regulations**, such as the EU's AI act proposal, impacting how algorithms can work with data and which use cases are acceptable.

The EU reacted to the explosion of data and the AI-driven revolution in data exploitation by introducing new laws. The following section looks at the two EU's landmark regulations: the General Data Protection Regulation (GDPR) and the new EU proposal for an AI act. Many other countries follow this model, at least for GDPR.

In contrast, many relevant regulations in the US are a reinterpretation of existing laws and rules in the light of AI. Thus, as a third example, we look at the Federal Trade Commission's approach of applying existing principles for the AI context.

A remark before looking at the regulations: Readers should keep in mind that regulations do not only reflect the academic discourse on AI ethics. Regulations also want to foster innovation and ensure that a country's economy can compete globally.

Data Privacy Laws: The GDPR Example

Today's worldwide gold standard for data privacy is the European Union's General Data Protection Regulation (GDPR) from May 2018. The GDPR is so well known (and feared) around the globe for combining, first, fines of up to 20 million euros or 4% of a company's global turnover, with, second, its global applicability if – simply speaking – companies process EU citizens' data. It does not matter where the company has its domicile or where they store and process such data.

The GDPR defines seven principles for data processing – obligations and restrictions companies and their AI organizations must follow. They are the easiest (and least expensive) to implement when adhered to in the design phase for an AI organization or concrete software solutions. "**Privacy by design**" or "privacy by default" are terms reflecting the ambition to consider privacy right from the beginning instead of trying to get it somehow somewhere into the code late.

The first of the seven principles demands **lawfulness, fairness, and transparency**. Any processing requires a legal base such as customer consent, legal obligations, or a necessity. Fairness demands not to misguide users, for example, with whom they are in contact or what they approve and accept.

The second principle is **purpose limitation**. When companies collect data, they have to be clear and transparent about the purpose they want to process the data for later. The stated purpose restricts and limits future data processing and usage for training AI models (if there is no other lawful base). Third, **data minimization** requires collecting and storing only data needed for the defined purpose. **Accuracy** demands personal data to be correct and not misleading – and to correct it otherwise. **Storage limitation** is about the duration of the data storage. Companies must not store personal data longer than needed.

The **security principle** (also termed integrity and confidentiality) requires that you secure your data. IT security organizations handle this in many companies. Furthermore, a later chapter looks at how to secure training data and environments and trained AI models.

Finally, there is the **accountability** principle, which requires you to take responsibility for handling personal data.

AI organizations must follow these principles, even if they generate extra work. In particular, AI organizations must understand where data comes from and for which purpose data scientists can use it (data lineage). Moreover, just stating to adhere to these principles is a time bomb, even if AI organizations really do. They need documentation as **evidence for audits**.

GDPR states further obligations. For example, a company has to confirm to natural persons on request whether they process any data relating to them and to provide a copy of all this data. Natural persons also have the right to get additional information such as how long a company stores their data, where the data comes from, plus the recipients of the data, or the purpose of processing (**data subject access request**). Companies need to have a comprehensive **catalog** of all applications, processing, and external data transfers. You cannot provide a requestor all his personal data if you do not know where you store which data. There is also the need for **Data Protection Impact Assessments** for high-risk data processing, for example, when data processing involves people tracking and surveillance or sensitive personal data. Thus AI organizations have to make sure that they follow company guidelines laid out by the IT department, the data protection officer, or legal & compliance teams to be ready for the next audit – or the next request from a data subject.

The EU's "AI Act" Proposal

On April 21st in 2021, the European Commission released a **proposal** for an Artificial Intelligence Act to foster the adoption of AI while addressing the risks associated with the new technology. The proposed legislation regulates AI algorithms broadly, from statistical approaches, over logic or knowledge-based systems, including expert systems, to supervised or unsupervised learning, deep learning, and other methodologies. The proposed AI act applies to AI systems within the **EU's jurisdiction and** AI systems **outside** if their output is used within the EU – and to companies, public agencies, and governmental offices, but excluding systems for military use. Violations of the AI act can result in **fines** of up to 6% of a company's global revenues, compared to "only" 4% in the case of GDPR violations.

The proposed regulation categorizes AI systems into four categories: systems with unacceptable risks, with high risks, with limited risks, and systems with minimal risks only (Figure 4-4).

Figure 4-4. *The AI Act's Risk Pyramid*

The planned AI regulation lists forbidden AI use cases considered as **unacceptable risks**. These are, first, systems using subliminal techniques or exploiting specific vulnerabilities of individuals resulting in harming the person herself or others physically or psychologically. Second, the proposed act forbids AI systems for social scoring by or on behalf of public authorities using data collected in different contexts or if the scores have a disproportional impact on individuals. Third, real-time remote biometric information systems in publicly accessible spaces are forbidden unless in the context of very specific crimes.

High-risk AI systems form the second category. Safety and health-related AI systems and components fall into this category, for example, systems for cars and trains or for medical devices. Also, AI systems for the following eight usages:

- Real-time or deferred biometric-based identification of persons

- Managing and operating critical infrastructures such as water supply

- Education and vocational training use cases deciding about the admission and assignment to educational institutions

- Recruitment processes, promotion or termination decisions, task allocation, and employee performance evaluation

- Decision-making about public assistance benefits, for dispatching and prioritizing emergency first response services, and for credit scoring

- Certain law enforcement use cases such as predicting whether persons will do an offense (again) or detecting the emotional state of a person

- Migration, asylum, and border control management, including polygraph use cases, assessing whether persons pose security risks or verifying the authenticity of documents

- Administration of justice and democratic processes with use cases such as researching facts and the law plus applying the law to case-specific facts

The proposed EU regulation does not forbid high-risk systems. Instead, the regulation lays out **specific requirements** for them. Companies have to identify and evaluate known and foreseeable risks in accordance with the intended usage purpose, thereby taking reasonably foreseeable misuses into account. A one-time assessment when developing or setting up such systems is not sufficient. Companies and their AI organizations have to continuously manage these risks over the system's lifecycle, considering insights gained during the production usage of the models.

There are additional detailed requirements for high-risk AI systems such as technical documentation, logging critical events, communicating the system's accuracy or robustness, and allowing for and ensuring human oversight.

The third category contains systems with **limited risks**. They have to follow specific transparency requirements. AI systems interacting with humans must unveil their non-human nature. When systems incorporate emotion recognition or biometric categorization systems, they have to inform the users. Finally, any system manipulating images, audio, and video content ("deep fakes") must make this transparent.

The **minimal risk AI systems** form the final fourth category. According to the EU, the majority of today's use cases fall into this category. The act does not regulate them but proposes establishing codes of conduct for them.

The proposed AI act results from a more than three-year process. Still, additional changes are likely until the act comes into force. Furthermore, the EU works on additional regulations impacting AI organizations, for example, related to liability questions. Most probably, other countries will put similar laws in place in the years to come, with this proposed EU AI act potentially having a significant impact and acting as a spiritus rector. Thus, AI organizations from outside the EU benefit from understanding these EU concepts, too.

The Federal Trade Commission's Approach in the US

The Federal Trade Commission (FTC) bases its regulation of AI systems on existing laws. The US agency's responsibility covers enforcing antitrust laws and fostering customer protection by preventing unfair, deceptive, or anti-competitive practices. The agency understands AI as a new technical approach for automated decision-making. So, the FTC does not focus on algorithm types to decide whether it is a "relevant" AI system; the FTC focuses on whether a system makes automated decisions. It regulates AI systems based on existing, older laws: the Federal Trade Commission Act (Section 5), the Fair Credit Reporting Act (FCRA), and the Equal Credit Opportunity Act (ECOA). Based on them, the FTC developed rules and guidelines for AI systems. The most relevant ones are:

- Transparency about the use of AI. For example, imitating human behavior without telling customers can be illegal.

- Transparency about the data collection, that is, no secret or behind-the-customer's-back collection of data without consumers being aware.

- Transparency whether automated rules change the deal, for example, by automatically lowering credit lines based on spending patterns.

- Right of customers to get access to their data stored and correct the wrong information. Depending on the exact circumstances, companies are obliged to have up-to-date, accurate data about their customers.

- Ability to provide customers the most relevant influencing factors resulting in the company denying a particular customer a specific request. For scores, a company has to be able to explain the underlying concept and to name the four key factors that impact an individual's specific score most.

- No discrimination on what represents protected classes in the
  US: race, color, religion, national origin, sex, marital status,
  age, or whether someone gets public assistance. Ensuring non-
  discriminating models requires adequate input data plus verifying
  the resulting AI model. The discussion on AI ethics provides helpful
  background information.

Not all of the rules apply to each kind of customer or contract. However, they exemplify
how existing laws regulate new AI organizations – and can be quite a challenge. A data
scientist might have the perfect AI model for credit scoring beating everything the banks
have in place yet. However, if he cannot list the four main factors influencing an individual's
credit score, the bank must not use the model. Thus, when looking at these rules, it is not
surprising that the ability to understand and explain how AI models work recently got
much attention. The terms used for this trend are "explainability" and "explainable AI."

Explainable AI

The need to build explainable AI model for complex prediction or classification tasks is
a real challenge. Data scientists can build (deep) neural networks. Neural networks are
black boxes solving difficult prediction and classification challenges – but they are black
boxes. Trained models consist of hundreds or thousands of formulas and parameters.
Their behavior is, for humans, impossible to comprehend – a not-so-desirable option.
Alternatively, data scientists can rely on easy-to-understand models such as logistic
and linear regressions or random forest. However, they do not perform well for complex
AI tasks – another non-desirable option. The recent and ongoing work on the topic of
Explainable AI (XAI) wants to overcome this catch-22 challenge.

Scenarios for XAI

XAI wants to help with three **scenarios**:

- The model – and not the application of the model on new data – is
  the deliverable.

- Justifications and documentations of model behavior for ethical and
  regulatory needs.

- Debugging and optimization.

All these scenarios require to understand how a trained AI model works.

It is well-known in object detection that neural networks do not always detect the actual object in the picture but react to its context. Even competition-winning neural networks have or had such flaws. In one case, the neural network did not recognize horses on images as everybody assumed. Instead, watermarks triggered this classification since only horse images in the training set had them. Suppose data scientists know – for example, in object recognition – which regions of a concrete image decide which type of object the neural network detects. Then, they can **debug and optimize** neural networks more efficiently. They detect more flaws similar to the watermark issue before models get into production and hit the real world.

The second XAI usage scenario reflects regulatory and ethical needs elaborated earlier in this chapter: checking for fairness and ensuring non-discriminating AI models. Customers and society do not accept the outcome of AI models as God's word: always correct and not to be questioned. Companies and organizations have to **justify** decisions, even if made by AI. As discussed above, the US Equal Credit Opportunity Act demands that banks communicate reasons when rejecting a credit. Otherwise, the bank must not use the AI model.

Third, the **model might be the core deliverable,** not its application on new data. A product manager wants to understand the buyers' and non-buyers' characteristics. The model has to tell him. It is not about applying the model to some customer data; it is about understanding what characterizes customers with high buying affinity based on the model.

The fourth, often-mentioned reason for the need for XAI **user trust**. The rationale: If humans understand how AI components work, they more likely trust the AI system's output. An XAI system for sentiment analysis could, for example, highlight the positive and the negative words and phrases to explain a text's sentiment rating (Figure 4-5). While it is intuitively clear that such additional information should foster user trust, studies proved the opposite. In one study, user confidence dropped when the AI component provided insights about its reasoning process, especially and shockingly when the AI component was sure to be correct and was actually right.

The conclusion is not to abandon the idea of boosting user trust with XAI. The conclusion is that projects have to validate this assumption empirically with real-life users. Humans often behave less rationally than we expect and wish.

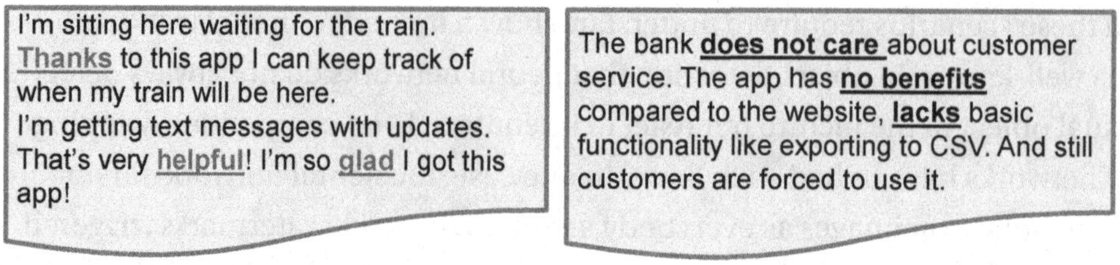

I'm sitting here waiting for the train. Thanks to this app I can keep track of when my train will be here. I'm getting text messages with updates. That's very helpful! I'm so glad I got this app!

The bank **does not care** about customer service. The app has **no benefits** compared to the website, **lacks** basic functionality like exporting to CSV. And still customers are forced to use it.

Figure 4-5. *Explainable AI for Sentiment Analysis – Highlighting Positive and Negative Words and Phrases in Sample Texts*

XAI is not a challenge for simple models such as linear and logistic regression. Everyone can comprehend the following formula and understand the impact of the parameters when looking at the estimation formula for monthly apartment rents in a specific city:

$$AppRent = 0.5 * sqm + 500 * NoOfBedrooms + 130 * NoOfBathrooms + 500$$

In contrast, neural networks predict and infer using hundreds or thousands of equations and weights. They deliver better results. However, any approach to presenting and understanding them in their complete baroque grandeur and complexity must fail. Thus, explainable AI chooses one lens to highlight one aspect of the neural network's behavior. Well-known are **two** of such **lenses**: global explainability and local explainability.

Local Explainability

Local explainability aims to understand the parameters which influence a single prediction. Why is this house so pricey? Because of the number of bathrooms or the garden size, or the pool? Local explainability does not aim to understand what – in general – impacts home prices in the market. It seeks to understand what influences the price of one single specific home most. One approach for local explainability is creating and analyzing a simple(r) and explainable model. This model describes the neural network behavior close (!) to the point of interest well but ignores the rest.

In more detail, the starting point is a neural network that approximates reality. Data scientists build the model, for example, using a data set with last year's home prices (Figure 4-6). The neural network might predict house prices well but is not human-comprehensible. When looking at the neural network and its weights, we do not understand what impacts the model most or whether small size variations matter for homes with 80 or 200 sqm (A and B in Figure 4-6). Thus, to achieve local explainability, data scientists perform three steps for each prediction value they aim to understand:

1. **Probing**. Data scientists need training data for creating an XAI model. Thus, they probe the neural network model around the points of interest. In Figure 4-6, they would look for the house prices the model predicts, for example, for 79 sqm, 80 sqm, 80.5 sqm, and 82sqm (case A), and for 185 sqm, 203 sqm, and 210 sqm (case B), respectively.

2. **Constructing an explainable model**. Data scientists use the data probes from step 1 to train a new model. This new model estimates the neural network behavior around A and B. The core idea is to use now a simple estimation function such as linear regression. The prediction quality is lower, but a linear function is easy to understand. Thus, in Figure 4-6, data scientists create two separate linear functions for the two points of interest, A and B.

3. **Understanding and explaining**. The linear approximation functions from step 2 allow understanding the local situations around A and B. Changes in the home size impact the price heavily for situation A, but not much for B. For models with more input features than just the living space size – garden size, number of bathrooms and carports, or whether the house has a balcony or an outdoor pool – the explanation gets, obviously, richer.

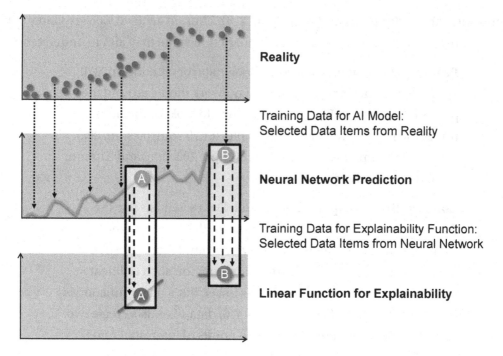

Figure 4-6. *Local Explainability Based on Approximating Model Behavior with Linear Functions Around the Points of Interest*

A remark for readers with a strong math background: It is essential not to calculate just the gradient but to probe around the points of interest. This approach reduces the impact of local disturbances and noise precisely at the point of interest that might not exist elsewhere in the nearby region (Figure 4-7).

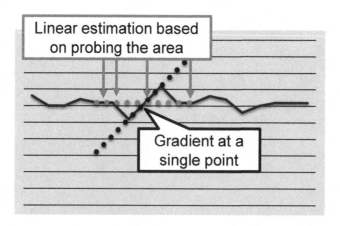

Figure 4-7. *Approximating the Local Behavior Using a Gradient vs. Linear Regression Based on Probes in the Area*

To conclude: local explainability identifies – for concrete predictions – which parameters to change for changing the prediction outcome most. It helps house owners to figure out for their individual houses whether adding a swimming pool or a car park increases the home values more.

Global Explainability

In contrast to local explainability, **global explainability** does not look at single predictions. It aims for the impossible: explaining an inherently too-complex-to-understand neural network and its overall behavior to humans. One well-known algorithm is permutation importance. Permutation importance determines the impact of the various input features on a given model's predictions.

The algorithm iterates over all input columns. Each iteration permutes the values of one particular column in the table (Figure 4-8, ❸). The more the accuracy drops, the more relevant is the permutated single feature. By comparing, for example, the accuracy after shuffling the sex and the zip columns, a data scientist understands that the sex has a higher impact on customers' shopping habits than the ZIP code (Figure 4-8, ❹).

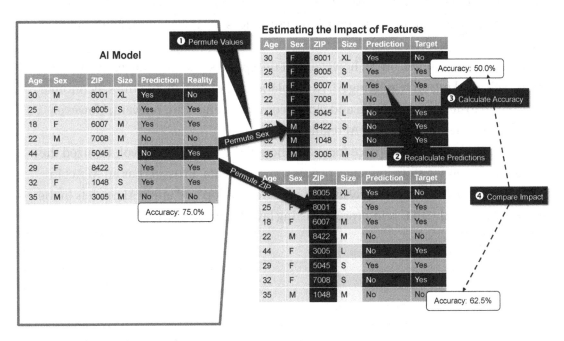

Figure 4-8. *Understanding Permutation Importance*

Explainable AI is a highly dynamic field with many new ideas coming up every year. However, due to regulatory pressure and societies expecting companies to behave ethically, commercial products today implement XAI functionality already as all the big cloud providers Google, Microsoft Azure, and Amazon's AWS proof. It is fascinating to see niche research becoming mainstream and relevant for many companies so quickly.

Summary

Three questions drive ethical discourse about AI usage in an enterprise context:

- Is the envisioned use case ethically acceptable? Should and do we want to do create and use this AI model?

- Do we engineer the model following ethical standards? Or do we, for example, use data in a way we should not just to get results faster?

- Is the AI model ethical? In particular, is the model bias-free and fair?

While bias-free AI models tend to align with business goals, other ethical aspects require balancing the ethical wishful with the commercial necessary – and having clear procedures in place about the responsibilities reduces the risk of bitter fights between teams and individuals.

The ethical discourse on AI, however, left the spheres of academia. Policymakers are actively working on regulating AI. AI regulation is not just about ethics. Economic aspects and fostering innovation and global competitiveness are relevant. However, ethical factors heavily influence AI regulation.

Data privacy laws are in place for years, though the EU's GDPR revolutionized the regulation of how we store, manage, process, and use data. Brussels has an AI Act in its pipeline, which will most likely impact how we develop and apply AI models in the future. In the US, the FTC reinterprets existing laws for the new world of AI.

The FTC requires, for example, that banks name the four key factors that result in declining a credit. A simple task in the traditional world with rule-based decision-making becomes a challenge if using neural networks with thousands of nodes and weights. It is no surprise that "Explainable AI" is such a hot topic with academia and the big cloud providers. Global explainability, for example, identifies the most important input factors for the overall model behavior. On the other side, local explainability looks at the factors influencing an individual prediction most. It is an eye-opener to observe how FTC's requirements, academic research, and big tech AI service innovations aim for the same: understanding AI model-based decision-making better.

CHAPTER 5

Building an AI Delivery Organization

When AI shall rock your company, a few projects here and there won't do the trick. You need a stable AI organization always ready to step in, help, and drive forward the transition towards a data and AI-driven company. This aim has implications for data scientists and the overall AI organizations. They have to widen their focus and shift it away from projects and models towards a kind of nation-building for AI within organizations. The task: build an AI organization that lasts.

Building an AI organization starts with identifying (internal) customers and securing funding. Customers mean that there are managers within your organization with a challenge they can and want to solve with AI. You have to convince them that your AI organization is technically competent, understands their business challenge, and helps them achieve their ambition. An AI organization cannot just focus on technology, or only within very few very large corporations with a high division of labor. From understanding the business problem to providing a solution for the exact context, this end-to-end perspective is the key to success with customers and for funding.

When new to corporate politics, it is especially crucial to be aware of two types of managers who support you. A *sponsor* helps you fund your service. She has a budget, or she can help you get funding from someone else. On the other hand, a *non-material patron* tells you how great you, your ideas, and your entrepreneurial spirit are. However, she has no money, and she does not lobby for any budget for you.

Once an AI organization ensured funding, the team must deliver the promised AI projects and sort out and run its AI infrastructure. Spending months on evaluating technologies and detailing all technical concepts seldom matches corporate expectations. Instead, sponsors and senior managers want to see that the AI organization delivers substantial business value.

© Klaus Haller 2022
K. Haller, *Managing AI in the Enterprise*, https://doi.org/10.1007/978-1-4842-7824-6_5

This chapter helps managers to succeed with these challenges by elaborating on three aspects:

- Shaping an AI service, that is, discussing typical elements of (AI) services and what choices exist.

- Looking at the organizational challenges of AI organizations running projects for creating superior AI models.

- Detailing the organizational needs of AI operations that run an AI model for years, for example, as part of the solution code or on a dedicated AI runtime server.

This topics combination helps managers prepare their service proposals and implement the actual service organization once senior management has approved the concept and ensures funding.

Shaping an AI Service

Personally, I always enjoy working with teams and managers to shape services within IT departments. You discuss a lot with engineers, managers, and customers, thereby deepening your knowledge about the topic and the organization. It is like Goethe's Faust, aiming to «… understand whatever binds the world's innermost core together, see all its workings, and its seeds.»

IT Services Characteristics

"Service" is one of those terms most people have more of an intuitive than a precise understanding. However, an accurate understanding helps to design AI (and other) services. Nearly twenty years ago, Joe Peppard elaborated on the main characteristics of IT services. They are: intangible nature, human factor, simultaneous provisioning and consumption, and no upfront demonstration.

The **intangible** nature of services reflects that services do not produce anything physical you can touch. Instead, a service is about someone doing something with something, which might change a physical or digital asset.

AI teams typically create, deploy, and maintain digital assets: AI models, inference results in the form of lists when applying and using the models, code for data cleaning and preparation, or interface configurations when integrating models in complex solutions. Additionally, AI teams might provide communication-intense consulting (e.g., helping internal customers formulate the question AI should solve) or support for technical incidents related to the model and its integration.

The **human factor** reflects that people interact and collaborate. Handling customer interactions professionally is as essential to avoid unhappy customers as successfully solving tasks and problems. Data scientists often have more academic merits than many of their customers. Still, they must patiently explain topics and communicate in an easy-to-understand language, value their internal customers' business know-how and experience, and show interest and understand the application context.

A second often overlooked aspect is the need for the customer side to contribute constructively. Suppose the integration between an AI runtime server and a sales campaign management system does not work. The owner or manager of the campaign management system cannot simply blame the AI team and refuse debugging on their side as well. Seniority and experience working with customers help handling such situations.

Simultaneous provisioning and consumption mean that service provisioning and consumption usually take place at the same time. If a physician examines me, I have to be present. The same applies to AI services. When AI models or interfaces fail, the AI team has to start working on the issue as soon as possible. Such incidents cannot wait, and such a support service cannot be prepared in advance and "place into stock." Depending on the exact request, the simultaneity is relaxed. Creating an AI model is a multi-week task. You might not want to generate a list in August with clients to target with Christmas tree coupons in early December. You have to do this in November or December, though one day or even a week earlier or later does not make a big difference.

The simultaneous nature has two implications for the AI organization. First, the management has to ensure that the AI team always has an adequate amount of work – not too much, not too little. Second, while proactive quality control for digital assets such as AI models is possible, proactive quality control is impossible for human interaction. Training data scientists on how to handle customers is an important measure. Still, when a data scientist yells at a customer, you cannot "quality control" the interaction before it impacts the customer. You can only do "damage control" afterward.

No upfront demonstration. It is usually impossible to demonstrate an IT service before the contract is signed and the service is established. You cannot test-drive an IT or AI service. An AI manager or data scientist can explain the sales team's success based on a new AI model. Still, such storytelling is no full equivalent to a live demonstration, for example, of the benefits AI can bring for a team optimizing an engine.

IT and AI service characteristics have clear **implications for an AI organization**. As mentioned, AI managers must carefully balance potential staffing needs and work gaps due to canceled or delayed projects due to the simultaneous provisioning and consumption of AI services. Managers with a consultancy or IT service provider background find this obvious. It is a new aspect for employees with a more academic background. The two other implications are similar – obvious for consultants, more surprising for others. AI team members require interpersonal skills for fruitful daily interactions, especially team members interacting with customers. Finally, when the AI organization requires identifying new projects, acquisition skills for AI managers are essential due to AI services' intangible and impossible-to-demonstrate-beforehand nature. However, the exact requirements and needs differ from project to project and service to service – and especially between companies with different corporate cultures and organizational structures.

AI Service Types

Creating an AI model for identifying the best city to build the next shopping mall vs. running and developing an AI component for a webshop suggesting customers the next best item to buy – these are two completely different services. One is an AI project (service); the other requires a stable AI operations service.

Projects – **AI projects** and others – have clearly defined goals and deliverables with set due dates and transparent effort estimations. Delivering "on time and in budget" is the aim of all project managers.

The core deliverable of an AI project is one (or more) AI model(s) complemented by secondary deliverables. One typical secondary deliverable is an inference result, such as a list for bank advisors as to which customers they should call to sell a new investment fund. The list gets created by applying the AI model to the customers' buying history. Another type of secondary deliverable is an analysis and explanation of an AI model pointing out and listing the (input) parameters with the highest impact on the overall result, for example, the carbon content or the exact heat when modeling steel production in a blast furnace.

AI projects need strong collaboration between AI specialists and employees from the line organization in the business to succeed. The latter have the required business know-how; they have to share it and must be willing to discuss and answer questions from the data scientists about the semantics of data or the exact business need and the business meaning of results.

The second service type consists of the **AI operations services**, which come in two variants. The first is integrating an AI component into a software solution. The software components trigger inference by the AI model whenever needed. There is no human involvement. The second variant of AI operational service requires data scientists or IT specialists to generate AI inference whenever needed or periodically, for example, every month. An operations engineer might generate a list of potential buyers of neon-orange handbags not only once but every month.

The AI organization has to balance the costs for automation with the costs for manual inference execution and handling customer requests – plus consider the operational risk of, once in a while, mixing up whom to send coupons for neon-orange handbags and whom for Ascot ties. Furthermore, even if the goal is complete automation, the second variant fits the minimum valuable product (MVP) concept: getting the solution to work and providing first results before working on advanced features such as full automation.

In any case, and in contrast to projects, AI operations services are a long-term commitment for many years. An AI operations service requires running and maintaining an AI platform, self-developed code, and interfaces and monitoring the model quality. From time to time, retraining the model is necessary, requiring short projects. Thus, an AI operations service needs (some) AI knowledge, but in general, running and maintaining an AI component is a standard IT operations topic. Data scientists are essential for model-related challenges. However, IT operations and application management specialists can take over all other tasks.

AI operations services have different collaboration patterns compared to project services. Projects require close interactions for efficiency. In contrast, operations services minimize interactions for cost reasons. They prefer asynchronous, highly structured, and standardized interactions: ticketing and workflow systems, not emails, phone calls, and in-person support. We discuss this topic in more detail later. Furthermore, Table 5-1 summarizes AI project and operations services' characteristics and compares them.

Table 5-1. *Comparing AI project services and AI operations services*

	AI Project Services	AI Operations Services
Goals and Deliverables	Create an AI model and either apply it once (or a few times) or analyze the model	Ensure that an AI team or an AI component continuously delivers AI inference results
Duration	Limited. Projects end once all deliverables are provided	Long-running, ongoing service, usually a few years
Duration	Single, defined project with clear deliverables,	Typically several years
Scope	• Create an AI model and • Apply the model on data once or a few times or • Analyze the insights the model provides.	• Frequent manual inference or • Running and maintaining an AI component with an AI model • Monitoring the AI model for degeneration • Trigger an AI model retraining
Interaction Pattern	Close collaboration between business and AI specialists	Mostly ticketing system

Characterizing AI Service Types

Is your AI projects and services organization similar to a fast-food or a fine dining experience? When you go to a fancy restaurant, a waiter will do some small talk, figure out what kind of food you prefer, and help you choose the menu. A sommelier pairs your meal with the perfect wine whereby considering your unique wine preferences. Then, the chef prepares the meal taking your allergies in mind. It is a different experience than ordering a Big Mac at McDonald's over a touch screen and getting a bag with your meal two minutes later over the counter. These are two different experiences, two different market segments, and two different price tags. Similarly, any AI department should be aware of what their customers expect and what the management funds.

Peppard's **IT Services Matrix** (Figure 5-1) is a simple tool to validate customer expectations and the AI organization's actual service proposition. Such a matrix considers two dimensions for assigning services to one of the four service types "service factory," "service shop," "service mall," and "service boutique."

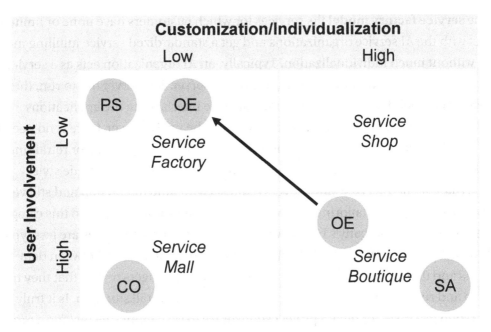

Figure 5-1. *Peppard's IT Services Matrix with AI Services: Platform Service (PS), AI Consulting (CO), Strategic Advisory (SA), and Operational Excellence Support (OE)*

The first discriminating criterion asks how much **customization** different customers need. The second criterion concerns the **interaction** between service employees and the user. How often and for how long do they communicate or collaborate to achieve a good result?

The "de-luxe" model is the **service boutique**: many interactions and much collaboration between the AI department and the customers and users, plus the option to place highly individual requests. Generating strategic insights for management advisory falls into this category. The top management has enough resources to fund dedicated AI projects – and they make sense to support high-impact strategy decisions. Also, unique product portfolio enhancement initiatives might request such services from the AI department. For example, a medical diagnostics company develops an AI-supported solution for diagnosing specific illnesses.

Besides C-level AI projects, regular AI projects for creating a new (non-trivial) AI model fall into the service boutique category. For example, data scientists have to analyze every detail when creating the first AI model supporting a marketing campaign. However, synergies might emerge later. Data scientists can reuse data sources or data preparation and cleaning logic when creating a second, third, or fourth model. It is a gradual development. Forward-looking AI managers foster and encourage such a trend.

The **service factory model** fits services for which customers have none or limited contact with the AI service organizations and get a standardized service fulfilling most needs without much individualization. Typically, an AI organization acts as a service factory when running AI models on an AI runtime server. The server has to run, the deployed AI models have to be operational, and the interfaces to the applications consuming the predictions and classifications must be up. However, there is no need for direct personal interaction with customers. Indeed, adding, improving, or retraining AI models means that the service switches (temporarily) to a service boutique style.

The category of a **service shop** covers requests with little interaction and still very much customer-specific tailoring. Software development projects fall into this category, so does designing a company's IT network infrastructure. However, they are less typical for AI organizations with their tendency to require much interaction between data scientists and domain or business specialists. When AI managers realize that they try to set up and run such a service, they should analyze the overall situation. Is it truly a service shop use case – or does a project sponsor try to save money by making a project look like a service shop instead of a service boutique project? Such ideas save money on paper, but tend to result in spectacular failures.

The **service mall** category provides services with low individualization or customization but a high interaction and involvement with the customer. AI departments operate in this category if they provide contractor-style internal AI consulting. In such a case, the AI organization has a pool of data scientists who help other teams with AI-related tasks. These data scientists need excellent social skills and the ability to tailor solutions for individual customers. There is much interaction with the customer, but the customization is limited. It is like renting a car. You can rent a car with two or four doors – you can either get an AI consultant with Python or with R know-how, respectively, if your team needs AI support.

The different models – service factory, service mall, service shop, and service boutique – are of particular interest to AI organizations. They have implications on which service attributes are essential for concrete AI services.

Understanding Service Attributes

A well-managed AI organization focuses its efforts on service attributes that matter. Philip and Hazlett's PCP attribute model helps AI organizations for getting priorities right. Their model distinguishes three types of service attributes: pivotal, core, and peripheral (Figure 5-2).

Most important are the **pivotal service attributes**. Pivotal elements for AI services and projects are typically the AI model and its evaluation and usage in the customer context. A concrete example is an AI model for churn prediction for a phone company, together with a concrete list of customers potentially terminating their contract next week. Such deliverables require AI know-how, including statistics and machine learning know-how, experience with tools and with bringing together all relevant data for superior models.

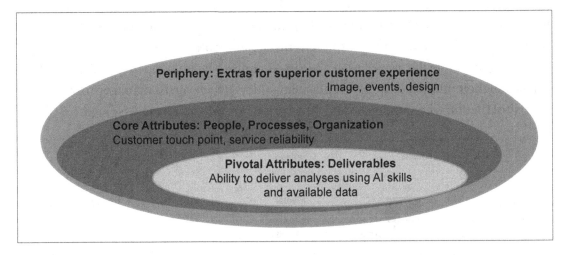

Figure 5-2. *Philip and Hazlett's PCP Attribute Model for Services*

Core service attributes are essential for long-term customer satisfaction. They cover people, processes, and organization and how these elements interact and create a service experience. They should ensure a reliable and friendly service. For example, how do customer touchpoints look like and how do customers accept them? Is there phone support, and can the hotline be reached as announced? Are there ticketing tools that are convenient to use and safe money compared to phone support? Can the project organization deliver on time? In various IT departments in entirely different areas, I observed that failing to manage such core service attributes results in escalations and frustration.

Finally, there are **peripheral service attributes** for the wow-effect. They excite customers, managers, and users. For example, are reports and lists designed well? Do team members present at conferences and are, thus, interesting to collaborate with in a project? Are there community events to foster relationships between the AI service team and the customers – and between customers? They excite and get you some goodwill when things go wrong once, but they are useless if the AI organization fails to deliver pivotal and core attributes adequately.

115

Understanding, categorizing, and designing service attributes is an important task when setting up an AI service. AI managers must understand and/or define them. Staffing, resource allocation, and management attention must reflect the importance of the various service attributes.

Designing (for) and Measuring Service Quality

You work hard and go the extra mile. Still, your customer, boss, or a friend of yours is dissatisfied with the result and expresses his or her emotions brutally direct. It is frustrating in private life as well as in a business context. It can happen as well with AI teams and their internal customers and stakeholders. However, AI managers can reduce the risk of friction when understanding the various service quality layers (Figure 5-3).

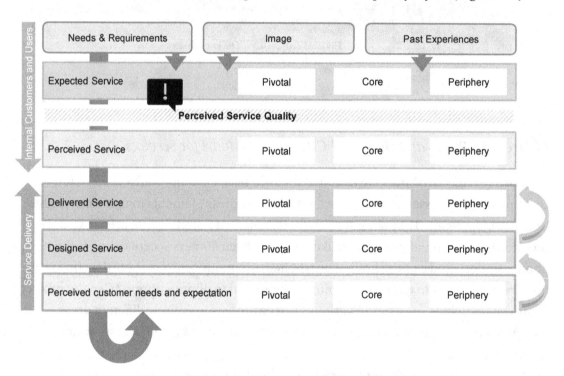

Figure 5-3. *Designing and Understanding Service Quality*

Customer satisfaction for AI (and non-AI) services depends on the perceived service quality regarding

- *what* the service delivers (pivotal elements) and

- *how* it is delivered (core and periphery elements)

Indeed, the **expectations** depend on **objective needs** and requirements, for example, to boost sales for a specific mobile subscription. The image and reputation of the AI team impact the expectations, so do experiences from previous interactions. These are **subjective influence factors**. Customers and users form an expectation, based on these various factors, what they expect from the service. They compare the *expected* service with the one they *perceive* being delivered. Customer satisfaction does not (only) depend on how well the AI team delivers a service based on *their* standards and guidelines. It depends on customer expectations and customer perception, too.

On the service delivery side, there are three different **service quality layers**. The starting point is the *perception* of customer needs and wishes. The next step is to *design* a service that meets these customers' needs. Then, the designed service is implemented and *delivered* by the AI team. Misunderstandings can happen in each step, impacting the overall adequacy of the service to meet customer expectations. As mentioned before, there is the customer perception of the delivered service, too. And, as always, some customers see a half-full glass, others a half-empty one.

The implication of this layered model based on various decades-old research on service quality from Grönsroos, Parasuraman, and Philip and Hazlett are threefold:

1. A stringent service design and experienced data scientists creating high-performing AI models are crucial.

2. Excellent PowerPoint slides about how the service works are not sufficient. An AI manager, an AI translator, a team lead, or consultants must work together with the delivery team to implement the concepts in a working organization. They have to check whether the service concept works in reality.

3. Managing stakeholder and customer expectations is a crucial part of the job for AI managers. Gaps are easier to handle before a project has started and while it is in progress than after burning all the budget.

These three guiding principles sound familiar, trivial, and evident for seasoned IT service managers and professionals with an IT or management consulting background. However, they help highly technical experts being in a lead role the first time and helps them make all painful experiences themselves. Running a service is a combination of communication and commitment. Based on my service design experience, writing down the key points of the service for potential customers helps. What do you do, and what are the limitations? It does not make additional communications obsolete. It fosters discussing needs and expectations – and prevents wrong assumptions on the customer side.

Managing AI Project Services

AI projects train AI models and derive actionable insights such as a target list of potential buyers for sales or typical buyers' characteristics for product managers. Managing an organization running many such projects requires awareness of the distinct workload patterns, understanding cost structures for preparing budgets, and communicating the results to the customers. First of all, however, an AI organization has to understand the needed capabilities.

The Capabilities Triumvirate for AI Project Services

The term "capability" comes from enterprise architecture. It helps to express what an organizational unit can do. It does not mean that an organizational unit has just the theoretical knowledge to take over tasks. Having a capability means that skilled staff, tools, process definitions, and methodologies are in place. The organizational unit, a team, or the AI organization must be able to perform or performs already the tasks successfully today.

Following the CRISP-DM methodology and going through all the stages for creating an AI model (i.e., without the deployment phase that is more an operations and less of a project task), we can identify all capabilities an AI project services organization needs (Figure 5-4).

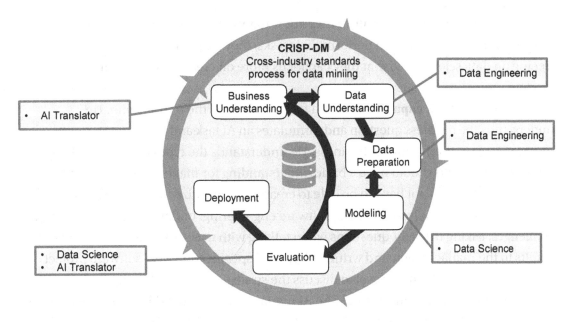

Figure 5-4. *Capabilities – A CRISP-DM-Centric Perspective*

The **core capability** is **data science** know-how. It is the base for creating an AI model in the "modeling" phase and assessing the AI model quality in the "evaluation" phase. It is the most apparent capability for AI organizations. Often underestimated are the two additionally needed capabilities of "AI translator" and "data engineering," though they require more time and effort in nearly all projects.

AI organizations need the **data engineering capability** for the "data understanding" and "data preparation" phases for performing the following tasks:

- Organizational challenges such as getting permissions of data owners to use their data

- The data transport: users and permissions, network bandwidth, transfer tools

- Understanding the data model and the semantics of the data

- Analyzing the data quality

The distinction between data engineering and data science capabilities is essential for staffing. Superheroes that can do everything are rare and expensive. Most companies have specialists who know SQL and one or more of the company's databases. They are highly efficient in preparing data exports. They can take over data cleansing and preparation without in-depth math, machine learning, or AI know-how. They are helpful in each AI organization, especially if they know data sources well. While they need to interact with data scientists to perform their tasks, they help the AI organizations by freeing up data scientists so that they can focus more on creating and optimizing AI models.

The third and last **capability** needed for AI projects is the AI translator. He understands the business question and formulates an AI task and presents the results to and discusses them with the business. He understands the types of insights AI models provide and should have a basic understanding for making some rough effort estimations. He does not have to be able to create and optimize models. The role is similar to business analysts in classic software engineering: understanding commercial challenges, asking the right questions when talking with users and customers, presenting results to the management, and writing high-level specifications. No software project sends its Java gurus to the C-level to discuss the strategic goals for the software development project. Neither should an AI manager send highly skilled and specialized data scientists.

Data science, data engineering, and AI translator form the capabilities triumvirate needed to support strategic decisions in an organization with one-time-use AI models. Figure 5-5 puts them in a broader context. Data engineering is close to the data providers and data sources. They ensure the AI organization can integrate as much of the data stored in any of the company's databases. The AI translators work with managers from various departments to identify which strategic decision-making can benefit from AI. In other words, creating which AI models could help analyze strategic and operational challenges by predicting individual or group behavior?

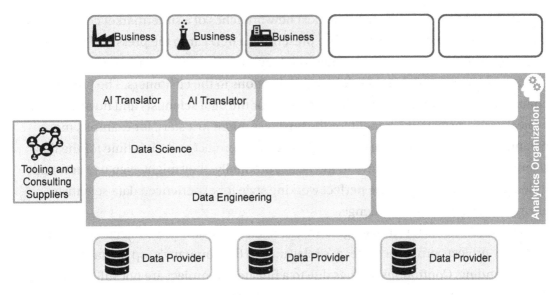

Figure 5-5. *Capabilities for AI Service Teams with Focus on AI for Advisory (White Boxes: Placeholders Needed Later for Other AI Services)*

Data scientists work with both of them, with AI translators to identify which AI models benefit the business and with data engineers to get the data they need. Indeed, the picture is not complete without external providers. For example, consulting companies might take over complete tasks or projects; contractors can augment internal teams. Plus, there are tool vendors (including public cloud providers) which the organization uses to improve its efficiency.

To prevent misunderstandings: Figure 5-5 lists many capabilities but does not imply that AI organizations should hire 2–3 persons for each of them. It is more like a list of tasks to be assigned to the team members. In a big AI organization, you and five other data scientists might work in a dedicated data science team. If the AI organization can fund two data scientists or engineers, the two together are responsible for everything. So, any team development strategy or process for hiring new specialists should consider all capabilities needed in general and the current team's strengths and weaknesses.

Workload Pattern

When an AI organization runs (mostly) AI-based advisory projects, the workload has a specific pattern. It is typical for IT or management consultancy but potentially new to AI managers and data scientists with a different professional background. Being in the **project business** means acquiring and delivering project after project after project.

Where should the company open the next new branches of a supermarket chain? Can we improve an engine the last time before presenting a new car to the press and the media? The projects follow the CRISP-DM methodology. Within a few days or months, the data scientists create a model and explain its implications to the customers. The project comes to an end. There are no follow-up tasks for the data scientists and no additional project costs. Instead, the benefit for the business might just start, for example, more revenues or lower costs (Figure 5-6, left). When one project ends, it is time for the data scientists and engineers to move to the next. Potentially, they might even use new and different technologies. It is the perfect working style for experienced data scientists always looking for a new challenge.

The implications for AI managers are the need to acquire new projects constantly. Only new projects prove the AI team's relevance to the company's top management and ensure funding. Continuous savings due to a finished AI project are not sufficient for an AI organization to stay relevant. If there is no next project, the CIO can shut down the AI organization without impacting the organization.

An AI organization (or a part of it) usually has a few or many parallel projects because there are always new ones starting, whereas older ones come to an end. Figure 5-6 (right) illustrates this aspect.

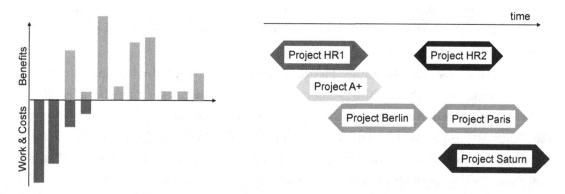

Figure 5-6. *Spending and Return Pattern (Left) and a Work Effort Structure for AI Organizations Focusing on Consulting-Style AI-Based Advisory Projects (Right)*

Budgets and Costs

Impressive titles alone, such as Head of AI, are insufficient for AI managers to establish a flourishing AI team with data scientists, data engineers, and AI translators. They need a budget to fund their team and pay for the tooling and infrastructure of their AI projects. For budgeting, they have to understand the cost structure of AI project services. We look first at an individual AI project's cost structure. Afterward, we broaden the view to the overall AI organization's one.

Two **basic cost dimensions** are essential for the understanding: inevitable costs vs. costs-by-design and fixed vs. variable costs. Inevitable costs are costs you cannot avoid. Costs-by-design means that the management makes an explicit decision where the organization invests more than the bare minimum. Two or three screens per data scientist workstation instead of one is an illustrative example.

The second dimension distinguishes fixed and variable costs. When you buy servers for machine learning, these are fixed costs. They have to be paid whether or not AI model training takes place. In contrast, on-demand-charged VMs in the cloud is an example of variable costs.

AI projects have the following cost structure consisting of four categories (as shown in Figure 5-7):

- Staffing costs for employees and contractors

- Infrastructure such as workplace for staff and compute and storage for machine learning training

- License fees for AI platforms such as SAP or SAS

- Centralized functions

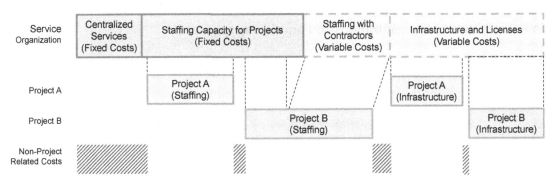

Figure 5-7. *Cost Structure for AI Project Services*

When working with internal employees, the **staffing costs** are fixed costs for the AI *organization*. The company has to pay their employees even if they have not enough work or them. When looking at *project* costs, staffing costs are variable. Projects typically only pay internally for employees to the extent they ask for their support. A mismatch between the overall work in projects and available staff in the AI organization is a challenge. Too much work causes project delays due to missing resources. Not enough work can result in cost-cutting activities. Finding the right balance is a crucial task for AI managers. One option to deal with workload fluctuations is to varabilize parts of the staffing costs by augmenting internal employees with contractors that scale up and down more quickly.

Infrastructure costs (besides employee workplaces) and **license costs** are fully variable for AI projects and AI organizations if the projects use cloud services. Alternatively, suppose an AI organization builds its own AI platform for its project. In that case, these costs are part of the **centralized service costs**. The latter also covers dedicated ticketing systems for interacting with customers or a Sharepoint to store documents. Therefore, project-focused AI organizations should carefully control costs for centralized services. If they are fixed costs, they can be a financial burden making the internal service potentially more expensive than externally sourced consulting services.

Once AI managers understand their cost structure, they have to ensure adequate **funding**. Details vary from company to company, and it can be a highly political issue. I just want to briefly explain two typical patterns: internal service providers and **prioritizing sponsors** (Figure 5-8). The last one is the most convenient. A senior manager (or some of them) provides the funding. The consequence: He selects the projects on which the AI organization works. He who pays the piper calls the tune. If he is in marketing, the AI organization most probably focuses on marketing and sales topics. If the sponsor is the head of an R&D department working on submarines, the AI team will optimize submarine-related challenges. Figure 5-8 illustrates a typical interplay between AI organizations and prioritizing sponsors.

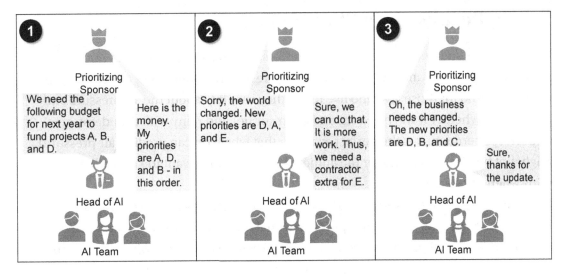

Figure 5-8. *Illustration of an AI Organization with a Prioritizing Sponsor*

The second option is the internal **service provider budgeting** pattern (Figure 5-9). It is typical for shared service organizations. The AI managers have to find customers in the organization who fund (at least) partially the costs. Variable, project-specific costs are easy to charge, whereas financing the fixed costs can be challenging. Adding a certain percentage to the variable costs to distribute the fixed costs is a typical mechanism.

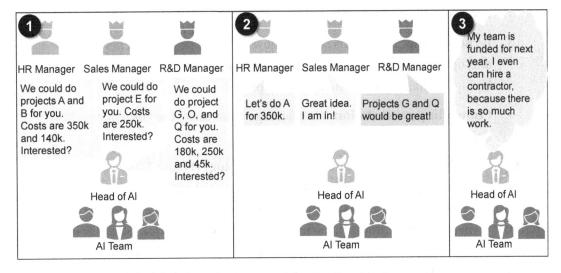

Figure 5-9. *Funding for the Service Provider Budgeting Pattern*

With this brief introduction, AI managers can discuss with their line managers and with their customers about costs and funding without needing a business degree.

Selling Results: Data Story Telling

Reaching your audience is vital for book authors, journalists, comedians – and AI organizations. Unfortunately, today's air pressure in Guadeloupe is of no relevance for most of us. And for those few who live there, they care less about the air pressure and more about whether it will be warm and sunny or hot and rainy with thunderstorms tomorrow afternoon. It is the perfect example that what you can predict (air pressure) might not be directly helpful for the audience. Instead, they want to know whether it rains where they are tomorrow.

Similarly, data scientists have to know their audience as well. They are experts in AI and machine learning algorithms, predictive analytics, and statistics to find unseen and unexpected correlations in enormous data collections. However, project results must meet four criteria, especially for projects helping with strategic decisions:

1. **Relevant** for the audience. The marketing department is interested in customer insights and sales, not so much in optimizing support functions.

2. **New insights** for the audience. What do they know after the AI project presented its results? What is new? A prediction that there is no snowfall in the Sahara desert tomorrow at noon might not be considered a groundbreaking new insight.

3. Be **actionable**. The results should allow for some improvements and optimization or efficiency gains. Obviously, the benefits must be higher than the investments in the project.

4. **Compelling presentation**. The message must make the audience curious and interested.

These tasks fit nicely into the CRISP-DM methodology (Figure 5-10). Checking the relevance of a topic is part of the business understanding phase when planning what to do and is repeated in the evaluation phase. Before presenting the results, AI project teams must check whether they provide relevant new insights and whether they are actionable. Certainly, presentations and written documentation should fascinate the audience as well. All these activities are highly creative and analytical. They are perfect tasks for AI translators who understand the customers and users and their business context in-depth.

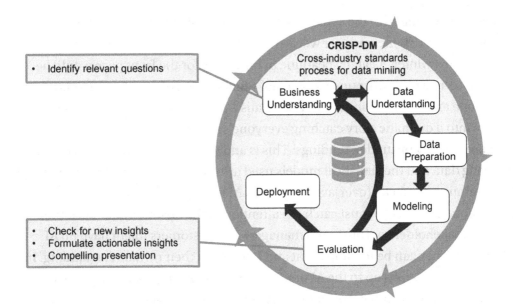

Figure 5-10. *How Data Scientists Get Heard by the Business*

I want to illustrate this explanation based on a concrete example of turning a model into an awaking story. A Swiss newspaper in July 2020 presented a perfect example of how to present data in a dull and meaningless fashion. They reported that the Federal Office of Public Health (BAG) announced 108 new Corona infections in Switzerland and Liechtenstein. These are more than twice as many as the previous day. There were 43 newly confirmed cases on Monday, 99 on Sunday, and 110 on Saturday.

When going through the four criteria, we see that the news fulfills the first criterion. The piece of information is relevant for readers in Switzerland. But does it provide new insights? No. Readers do not know afterward whether the numbers are comforting or alarming. They heard for weeks and months that the numbers from one day to the next could not be compared directly, because they are higher on certain weekdays. Thus, a piece of news just providing today's number and the information from three previous days is worthless. The number for each weekday from last week gives a much better indication. This Tuesday, we have 108 cases; one week ago, only 70. Yesterday, we had 43, the week before 63. While this might not sound alarming, this is different if the seven-day average indicates an increase of 12% per week.

Still, this information is too abstract for most readers to be relevant for them. The absolute numbers are too low (726 out of 8 million). It is more likely that you get 5 + 1 correct with the Euromillions lottery. However, the yellow press would make it relevant for everyone in Switzerland by connecting the numbers with the traumatic first lockdown in March of the same year. The number of daily infections on that day was 1063. With a 12% increase per week, Switzerland would have reached the same level again next winter. A catchy headline in big letters for the front page would be "2nd lockdown expected around Christmas." Suddenly, an abstract number is relevant for everyone and even actionable for politicians – a perfect example of turning simple numbers into a dramatic story catching everyone's attention.

To prevent any misunderstandings: This is and was not an actual forecast. The underlying data and the statistical models used are inadequate. The example highlights what communication can (over)achieve. When AI managers want to innovate the organization, their results must catch the attention of the business and reach their audience. Stakeholders or (business) managers and customers might have no AI background. They can be too impatient and busy to get their own understanding. In such (typical) cases, it is up to the AI organization to let AI translators help out or ensure that some data scientists have outstanding communication skills.

Managing AI Operations Services

When AI organizations create models and integrate and run them in larger solutions, the set-up and working style of (this part of) the AI organization evolves further. Capabilities, financials, skills, and cost-drivers differ from a consulting-style project-focused AI organization. For AI operations services, new topics emerge, such as target operating models and clear support touchpoints to the outside world or model management. The AI organization needs six instead of just three capabilities.

The Six AI Capabilities

When looking at the needed capabilities for AI operations services, there are core tasks for an operations service – and tasks an AI operations organization has to perform, too. Indeed, it has to be able to run at least small-scale AI projects, such as creating (less complex) initial AI models and retraining models. Thus, an AI operations organization also needs the AI projects capabilities triumvirate: data science, data engineering, and AI translator. (Figure 5-11).

When looking at the additional capabilities, especially relevant for AI operations organizations is the **integration engineering capability**. It covers installing and patching software components and configuring interfaces the AI organization takes care of afterward. Integration engineering includes connecting the AI components' interfaces with other solutions to enrich traditional business processes with AI logic. An excellent example in the AI context is installing software such as RStudio Server or integrating RStudio with LDAP, which comprises typical **application management** (AM) and **IT operations** tasks: solving issues, installing patches, and coordinating releases with other applications and the underlying infrastructure. It might also cover integrating the AI components with the IT department's central monitoring solution, which looks for aborted processes or errors and warning messages in logs to detect issues, ideally before they impact users.

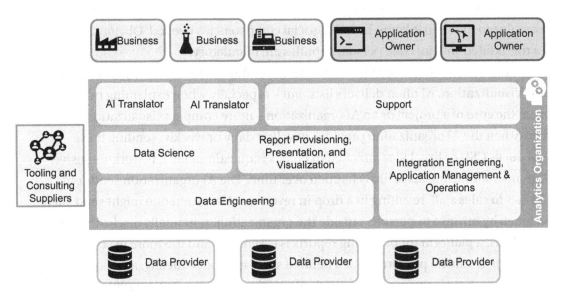

Figure 5-11. *Capabilities for AI Operations Services Organizations*

The **support capability** reflects that components do not always work as expected, requiring a support specialist to analyze the situation and fix issues. Thus, AI operations implies the need for a support team or at least for clearly defined specialists who take over user and customer interactions and ensure prompt reaction. They are the face to the users and customers, for example, if users miss a result, an interface breaks down, or interface configurations have to be changed. In addition, in an AI operations organization, they might perform routine AI tasks, for example, in the context of retraining models.

Involving application management and support specialists in routine AI tasks increases the AI organization's resilience. These specialists demand mature scripts and clear documents when they take over a model from AI projects and data scientists.

The benefit for the organization is obvious: From now on, when a data scientist gets sick, not everything breaks apart because he is not the only one anymore knowing all the scripts and tricks to run and retrain an essential AI model for online sales.

Based on my experience, one single person can take over support and application management in small and focused organizations with not more than a handful of user interactions per day and a few components and not-too-complex systems and interfaces. Still, these are separate tasks and capabilities. One focuses on reactive user interactions and solving easier challenges, including maloperations by end-users (support capability). The other, that is, application management, is more proactive maintenance and analyzing the trickier issues when the AI components or interfaces fail. The combination makes sense due to the similar know-how and social competencies needed. Obviously, however, if a role is filled with just one person, deputies are mandatory.

One capability might or might not be needed: **Report Provisioning, Presentation, and Visualization**. AI often delivers lists, but – especially when explaining models are at the core of a project or an AI organization – more complex visualizations help. Also, when the AI organization provides new lists daily or weekly, sending these lists out as CSV or Excel files attached to emails is questionable. It is work-intensive. Plus, operational mistakes likely happen over time. The AI organization sends out old lists to sales staff, resulting in a drop in revenues. Also, someone might send such an email by mistake to persons outside the organization, potentially disclosing sensitive data. Thus, a platform for providing reports is essential – and the more users, the more important is a stable platform with a defined support model. Ideally, an AI organization feeds its results into the company standard solution such as a Business Intelligence (BI) solution or an Enterprise Resource Planning platform.

Workload Pattern

Recurring business and flattening the (workload) curve are two slogans on how AI service organizations work with support and application maintenance tasks.

Recurring business means that the AI organization provides ongoing services. A support organization, software maintenance, and application management are necessary for users and customers of AI models or components. The AI model generates business benefits or operational efficiencies, but only as long as the AI service organization works. If the top management shuts down the AI organizations, the benefits and efficiencies are gone from that moment. Figure 5-12 illustrates this spendings and benefits pattern over time.

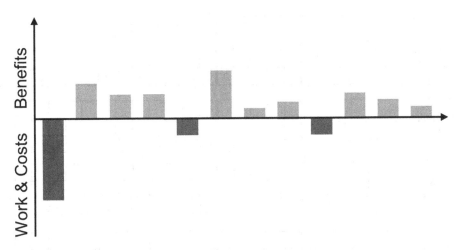

Figure 5-12. *Understanding Work Effort and Costs vs. Business Benefits for AI Operations Services*

The "flattening the workload curve" slogan reflects that operations services can, up to a certain degree, shift workloads to different days, weeks, or even months. Application managers and support staff can work for several teams and applications in parallel. The example in Figure 5-13 illustrates this. The AI service organization supports and maintains AI models and interfaces for three applications. Additionally, they run one project, which goes live soon. If there are separate teams for each of them, the App X team requires three specialists, App Z three and two for App R and one for Project 13. Altogether, the organization has to pay for nine specialists. When sharing, they need four – and if they manage to shift one workday from day three to another day, then only three. Many maintenance tasks can be moved by a few days or weeks, which is the key to flattening the workload curve and the organization's efficiency.

	Week 1					Week 2					Week 3					Week 4					Week 5				
Required Engineers for App X	0	0	1	0	0	0	1	0	0	0	0	0	0	3	0	2	0	1	0	0	0	0	0	1	2
Required Engineers for App Z	0	2	3	0	0	0	0	1	0	0	1	1	1	0	0	0	0	0	1	1	1	1	0	0	0
Required Engineers for App R	2	0	0	2	2	1	1	1	1	1	1	1	1	0	1	1	2	2	0	0	0	1	1	1	0
Required Engineers for Proj 13	1	1	0	0	0	0	0	0	1	1	0	1	1	0	0	0	0	0	0	1	0	0	1	1	0
Required Staff	3	3	4	2	2	1	2	2	2	2	2	3	3	3	1	3	2	3	1	2	1	2	2	3	2
Available Staff	4	4	4	4	4	2	2	2	2	2	2	3	3	3	2	3	3	3	3	3	3	3	4	4	4

Figure 5-13. *Workload Planning for Support and Application Management in a Multi-Application Setting with One Project*

Understanding and Managing Costs Drivers

The big challenge for managers of AI operations services is to invest sufficiently in preventive measures for stabilizing the solution and in the availability of support and engineering staff to handle incidents and support cases. The intentional investment into running the service does not generate profits. Not spending enough, however, results in service negligence costs and harms the business. Two radical approaches illustrate the challenge. The support desk can be one intern reading emails and helping with wrongly used GUIs on Monday morning, Wednesday, and Thursday afternoon. Alternatively, a 7/24 phone support hotline provides access to three senior engineers on call, plus professional monitoring and alarming, making incidents even less likely. The costs differ, so does the impact on the rest of the organization.

The **cost for running the service** is the money the organization (willingly) spends for the support capability and ongoing application management. Therefore, the first cost domain is **support and problem management** with the following three primary cost drivers (Figure 5-14):

- **Customer interaction design**. Phone, drop-in service-desk, email, ticketing systems, FAQs, one-page documentation leaflets, trainings – there are many ways to provide customer and user support. However, the right approach depends on the volume and complexity of the requests and impacts the time spent with users and customers.

- **Support and reaction times**. 7/24 support requires more engineers and generates higher costs than if the support only works from 9 am to noon on weekdays. Also, if tickets have to be answered within five minutes and not within four hours, staffing needs and costs explode.

- **Diversity of the service portfolios**. What technologies are in use, such as Python, SAS, SAP Analytics, or R? How different are the implementations and business aims of the implemented services? Even just helping customers use the GUI requires the support staff to understand the technologies. Much more technical know-how is needed when having to solve support cases requiring real engineering. The underlying question is whether one engineer per shift can handle all requests knowledge-wise – or whether there is a need for two or three engineers. It is a vital question before adding any new technology to the service portfolio.

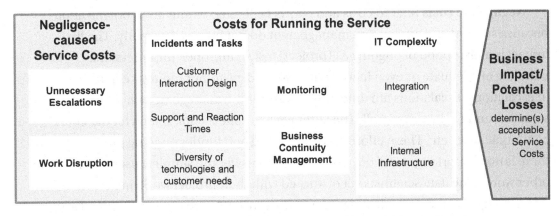

Figure 5-14. *Understanding Costs for Continuous Data Science Services*

The latter aspect reflects the impact of technology and technologies. The **complexity of the IT infrastructure** impacts the service costs as well. First, this covers more **integration**-related elements such as the number and stability of the interfaces. For example, is a messaging middleware in place, or is the communication and interaction self-coded? Are the applications the AI components work together stable, or are there always incidents and discussions which solution is the root cause for issues? These questions relate to the interplay between production systems. However, an AI organization must also maintain its **internal** hardware and software **infrastructure** with CI/CD pipelines, especially if integrated into pipelines from other software development teams, runtime servers, AI platforms, etc.

Business continuity management and **monitoring** represent additional cost blocks. Details depend on whether there is a solution and processes in place by the IT organization. So again, they do not help you generate business value; they "only" detect problems earlier or reduce the impact of issues.

All these mentioned costs for support and application management are directly visible. Controlling specialists understand precisely the money spent, but not the benefits. Such constellations are an invitation for budget cuts during cost-optimization initiatives. One person less? You should not see any impact. We could cut one engineer without any service impact; why not cut a second position as well? It is like the story told about a university. To save money, they reduced the room temperature in winter to lower the costs for heating. You can repeat this exercise year after year – until your electricity bills explode because your staff brings electric radiators once offices were too cold to work. It is similar to application management and support costs. They can be cut more or less to zero. The impact is severe: negligence costs within the AI organization and business impact outside the AI organization.

Negligence costs refer to costs and efforts generated within an AI organization because support and application management do not function properly. The first impact is a disruption of ongoing AI (project) tasks. Improper, unprofessional handling of adequate or even inadequate user and customer requests can result in escalations. Escalations are time-consuming for the management, but for data scientists and engineers as well. The latter have to collect data, search for old email communication, etc. These efforts have to be paid and hinder ongoing projects. In case there is no support organization, data scientists answer the simplest user questions. In other words: the data scientists get disturbed with their project work on new AI models if first-time users do not understand the GUI or push the wrong button. Also, if systems stop working, the time pressure for patching or reconfigurations is much higher. Thus, more persons have to work on the issue than performing measures preventively in a less hectic setting. Furthermore, every issue you detected with monitoring before users call or open tickets reduces the workload for the support channel.

Finally, the **business impact** and **potential losses** reflect the non-realized business opportunities or savings because an AI is not working. For example, suppose the AI component generates upselling and cross-selling recommendations accounting for 20% of the revenues of the webshop. In that case, these revenues are gone when the AI component is down or starts making bad suggestions. If the business managers are not aware, AI managers should bring this on the table in budget discussions when senior managers challenge the costs for model quality and service stability.

Model Management

"It is mine, I tell you. My own. My precious. Yes, my precious." The famous quote from "Lord of the Rings" is a seldom heard expression of care and obsession about an artifact. AI organizations do good taking similar care for their AI models and treating them as valuable corporate assets and intellectual property. That is the task of model management.

Model management has only limited real IT or AI challenges. Still, it is the most technical of all AI operations challenges. It covers three main aspects. The first relates to quality assurance during the CRISP-DM model engineering process. The second aspect is monitoring and managing the model quality after deployment. The third and last aspect is how the organization handles models as artifacts during their lifecycle (Figure 5-15).

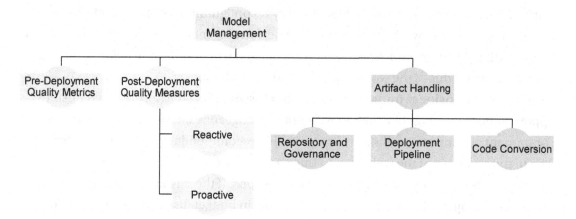

Figure 5-15. *Topics in Model Management*

A previous chapter was completely dedicated to quality assurance during the model creation with some brief remarks about post-deployment quality measures. We highlight a different perspective in the following.

A **reactive approach** is an **automated model scoring and KPI monitoring** looking for changes ("drift") in the input data and the derived predictions or classifications (Figure 5-16). Does the input data distribution change, or more concrete, for example, its average or lower and upper values? Does the output distribution change over time, or the business impact change as a drop in the conversion rate? When the monitoring detects that a KPI drops below a threshold, it can generate an alarm for the data scientists such that they are prompted to retrain the AI model. The simplest option is to train the same neural network architecture with the most up-to-date data (which might still might contain older data if still relevant and helpful). The most comprehensive retraining approach is to repeat the complete model development process, including feature development and hyperparameter optimization.

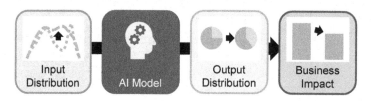

Figure 5-16. *Analytical Model, Neural Networks, and Drift*

It is a reactive approach since the retraining starts only if a detectable decay of a KPI signalizes a need. On the other hand, a **proactive** approach is to retrain even if the KPIs still look good. The **champion-challenger pattern** falls into this category. The model currently in production is the champion model. Though the model (still) meets the quality requirements, data scientists already prepare a new model named the challenger model. Data scientists rely on the latest data or try more sophisticated optimizations and compare the champion and challenger models' performances. If the challenger model outperforms the champion model, the challenger model becomes the new champion model. Otherwise, the data scientists discard the challenger model and start working on the next one. Figure 5-17 illustrates this champion-challenger race. In general, the champion-challenger pattern is a kind of A/B testing trying to find the best solution.

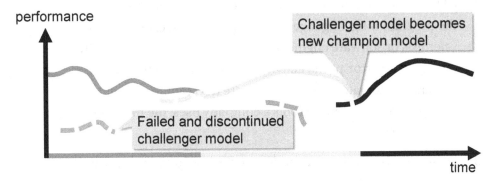

Figure 5-17. *Champion-Challenger Pattern*

Besides the quality aspect, looking at **models as artifacts** and optimizing their handling and, hopefully, smooth progress through corporate processes is another model management topic. Data scientists work with them, share them, document them, store and archive them, or version them. Many tasks are familiar from software engineering; others are new. However, three aspects are especially relevant (Figure 5-18):

1. Tool-based **managed deployment processes or pipelines** enforce corporate procedures and reduce operational risks and hurdles. Classic IT organizations tend to implement sophisticated approval processes for signing off requirements, quality assurance, and deployments using tools such as Jira or ServiceNow. This old-world approach focuses on auditability. It is mandatory for many organizations and potentially binding for AI organizations and how they create and deploy models. The

newer trend is that developers and data scientists automate the release building and deployment using CI/CD pipelines. They make manual work redundant. They save time and reduce the risk of deployment mistakes (wrong environment, wrong model, wrong parameters, etc.). AI managers should check whether their processes are fully integrated into the company's change management processes, rely on the same tooling, and fulfill the audit requirements.

2. **Code conversion** functionality helps data scientists and software developers when they work with different technologies. One of the options discussed in previous chapters for integrating an AI model into a software solution is to (re)code the model in the programming language the solution developers use. Data scientists or software engineers can perform this task manually, obviously an error-prone activity. Alternatively, code conversion packages automate this step. The more development languages and the more different AI tools a company uses, the more critical is this aspect.

3. Data scientists must store models as valuable corporate assets in a central **repository**. A repository guarantees that models do not get lost or deleted by mistake, enables collaboration between data scientists, and ensures auditability.

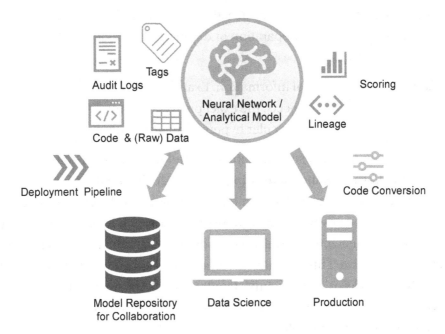

Figure 5-18. *Model Management in a Corporate Context*

The repository needs some additional explanation. AI organizations can, for example, use the IT department's code repository that they set up for software engineering. An AI organization would store, obviously, the AI model's parameters and hyperparameters. Furthermore, additional information is vital for the deployment process, for searching for models, or for surviving audits. Such potential additional information is:

- The data preparation and model training **code**, accompanied by the actual **data or a reference to the data**. Only this combination enables data scientists to recreate precisely the same model at a later time.

- **Data lineage** documenting how the data got from the source to the point it became training data. For example, does the financial data come from the enterprise data warehouse or an obscure Excel created by a trainee in the housekeeping department and copied and modified by five teams afterward?

- **Scores** measuring the model performance. They document the rigidness of the quality assurance activities and prove the adequacy of the model.

- **Audit logs** documenting crucial activities such as who changed the training code or data preparation or deployed the model to which environment.

- **Tags** storing additional information. Examples are the data scientists that worked on the projects, the exact purpose of a model, or why a data scientist used a particular hyperparameter value.

Storing AI models in a repository together with additional (meta-)data, enforcing a rigid deployment process, and monitoring the after-deployment performance of models – these are the three pillars of model management. Of course, model management is only a sandwich layer between the trendy and hype data science work and the glamorous management level. Still, model management is the glue needed for a sustainable AI initiative. This glue allows corporations to benefit from AI initiatives for years, not only for weeks or a few months. Finally, model management gives future data science projects a kick-start with all the model repository information and artifacts – reuse cannot get more comfortable.

Organizing an AI Organization

Congratulations! You shaped your service, analyzed the necessary capabilities, prepared a budget, got funding, and hired the right talent. However, to kick-start your AI service, your staff has to know how to work on which tasks, how to collaborate within the AI organization and the rest of the company, and how to handle and complete user and customer requests.

AI project services are forgiving if the AI organization is not super structured – they (should) have explicit goals. The AI management just assigns engineers and data scientists to projects and clarifies how much time they can spend on which project. The CRISP-DM methodology structures the daily work of the AI experts (Figure 5-19, left).

AI operations services require a more sophisticated organizational setup. Just assume that an online shop AI component suddenly proposes ski equipment to customers searching for beachwear. In that case, the AI organization is under high pressure to fix the issue – calmly, speedily, professionally.

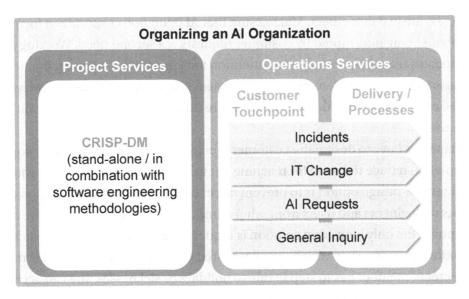

Figure 5-19. *Topics for Organizing an AI Organization*

The big difference to projects is that most operations organizations serve multiple (internal) customers, applications, or services in parallel. They react to incidents and dedicated customer requests. Typical tasks are restarting components, fixing interfaces problems, or configuring a new interface. They schedule preventive maintenance such as installing patches when vendors provide new ones. Such tasks keep engineers busy for a few hours or days, not for months or years. To keep the support specialists' work pipeline full, they take care of multiple applications and interfaces as illustrated in Figure 5-13 – or they work in the context of one super-complex application.

The challenge for service organizations: Everybody wants to benefit from the lower costs of a shared service, that is, specialists work not only for one interface or one small component. Still, everybody expects superior service, even if other customers and users have issues simultaneously, and compete for the attention of the support organization.

When preparing an AI operational team for such daily challenges, organizing the actual service delivery (such as implementing changes on systems) is just one task. Crucial for success is also the **customer touchpoint** as part of the core and peripheral service attributes. Here, users, customers, and the AI operations organization interact. Answering phone calls from first-time users or sending emails forward and backward to get all details for a service request is time-consuming and keeps AI operations specialists busy. Having to sit in meetings and workshops can be worse. No manager wants to pay 20 hours for a two-hour task due to time-consuming workshops, notes being made, and employees exchanging many emails. AI organizations have to prevent such situations.

But costs are just one aspect. Making sure that AI operations experts can focus on tasks without being interrupted by customer requests is a second need. Working on critical updates requires uninterrupted attention for some time. Flattening the workload over time is a third crucial element for cost efficiency. Non-urgent tasks are delayed to low times, even if customers think their needs are the company's most important and urgent ones.

AI managers have to design the customer touchpoint carefully. FAQ pages and good documentation reduce the need for reaching out to the AI operations team. A widely followed service-design option is to prevent direct contact between AI operations specialists or engineers and users from other components. Group mailboxes are a starting point; the only long-term solution is request forms or a ticket system such as Jira. They ensure transparency in which teams required how much support. Furthermore, if another team member has to step in due to holidays or sicknesses, she can access the information about all incidents. There are cases when users and customers need a specific contact person. In that case, this requires additional staffing in the AI organization – and someone has to fund such premium service.

The main challenge for the **service delivery** is to ensure a smooth processing and routing of requests and a good collaboration within the AI organization and with other teams. All involved specialists must know which activities they are carrying out, which interim results they receive from their colleagues, what they have to deliver as results themselves. The solution? Clearly defined and agreed-upon processes.

For an AI organization, four processes or process areas are of particular interest:

- Incident process: Something that worked before is suddenly not working anymore and has to be fixed. An example is an AI runtime server abruptly discontinuing to reply to requests.

- Change Request: Something that works should be changed or moved to the next level. For example, there should be an alarm for model degeneration. In the future, the alarming should take place when less than 20% of the customer buy proposed items instead of 30% now.

- AI Requests: questions, wishes, and requirements related to the actual AI models, for example, retraining a model or clarifying questions about the model.

- General inquiry: standard input window for other questions and not time-critical aspects.

A **process definition** clarifies who is doing what, when, and using which tools and systems. Managers and experts with a strong IT or science background tend to define processes that cover all possible cases. This approach is super-expensive, identifying every conceivable case is often challenging or impossible, and the process definition distracts the users from the core ideas, that is, what operations and support specialists should focus on to deliver an excellent service.

The preferable approach for defining processes is to focus on modeling and training "happy flows." These are the cases where the process goes through smoothly. Only frequently occurring error cases are modeled, not every exception. Exceptions are seldom an issue in practice for independently acting, well-trained support and operations specialists. They find a solution. If the situation is beyond their area of expertise and comfort, they contact the management. Just one warning, this assumption might not be correct for every company culture and each cultural context.

A **swim lane process model** (Figure 5-20) is a straightforward process modeling technique useful for documenting AI operational processes. Each swim lane represents a department, a team, or a specific role. A process consists of different steps. Their placement in a column or swim lane defines who is performing the task. If the employees work with a specific IT system to fulfill a particular task in the process, the system used should also be specified. Swim lane process models (in contrast to, e.g., UML Sate Machines) can only visualize processes without too many case distinctions or loops. If a swim lane model gets too complex, splitting it into two or three can help. Otherwise, the process is probably too complicated for everyday usage by a team. Once an AI organization has defined all its core processes (e.g., the four mentioned above) and trained all team members, the organization is ready to work.

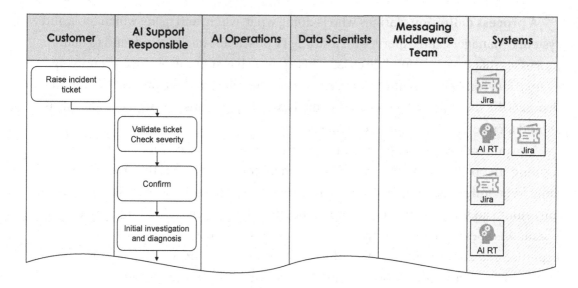

Figure 5-20. *Example of a Swim Lane Process Model with Five Actors and an Extra Row Indicating the Systems Used by the Actors*

A guiding principle for an AI organization designing and implementing its processes is **compatibility with the company's other processes**. If other teams have to perform a specific step, these teams must also agree on the processes and tools. In many cases, IT departments define their processes based on the de-facto standard, the **Information Technology Infrastructure Library** (**ITIL**). However, just knowing the standard does not help an AI organization much. IT departments tailor ITIL processes heavily. Who is precisely approving requests coming from which system using which forms? What are the tools for which procedures? The details are company-specific but crucial for working AI processes.

Many AI solutions come with integrated workflow engines helping an AI organization to structure its work. It is a convenient option, but AI organizations should clearly understand the limitations: they might have to use, in parallel, other systems and copy tickets from one system to another. Suppose a database team should export and hand over some data. Will they accept tasks assigned to them by a workflow system of the AI organization? Maybe, but for sure not in large IT departments. What happens when the marketing team notices that the "customer who buys dresses by also …" feature does not work? Will they open a ticket in the company-wide **incident management** system and let the IT department figure out the details? Or would they love to figure out themselves whether it is an AI or a webshop issue, search where the right incident ticketing system is, and open a ticket in this specific tool of the AI organization? Apparently not. AI organizations have to ensure that their tooling for processes and requests follows the company standard. Otherwise, they look for trouble when their systems have any relevance to the business.

While processes are crucial, many organizations nowadays talk about **(target) operating models** (TOMs) that define how an organization works. TOMs describe an intended future operational model. In contrast, the term "Current Operating Model" refers to the current status quo. An AI organization providing operational services usually needs a TOM. The purpose is not so much for keeping the operational processes up. The real purpose is more to discuss strategic directions for the AI organization, elaborate on how to transform the organization, and clarify its possibilities and limitations.

A (target) operating model cover the following core aspects (Figure 5-21):

- **Service pledge** with the value proposition. Here, the AI organization's business value and shaping the service as presented at the beginning of this chapter help.

- **Processes**: Which activities are processed in which order by whom and using which tools? We discussed this a few paragraphs up.

- **Staff and organization** taking over the tasks defined in the processes.

- **IT** with the infrastructure and applications (and their parameterization) as a core "enabler" for (efficient) process execution.

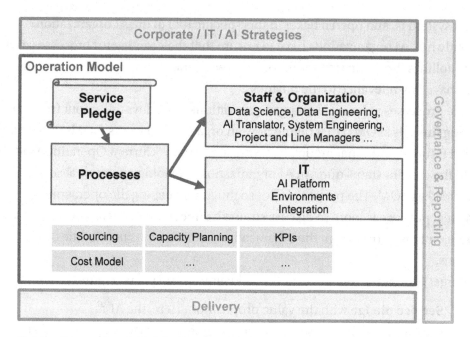

Figure 5-21. *Operating Models for AI Organizations*

AI managers can specify the first three topics and write them down quickly based on this book's input. Many supporting topics, including KPIs, capacity planning regarding IT resources, staff, or souring strategies (e.g., internal employees, contractors, consulting companies) require, for sure, more clarifications. The fourth important aspect of an operations model is the needed IT infrastructure. It is the topic of the following chapter.

Summary

This chapter elaborated on the specifics of managing an AI organization and not just any team within an IT department. In particular, it covered three topics: shaping an AI service, managing the AI project services for creating new AI models, and ensuring reliable operations when it comes to running and maintaining (the infrastructure) for AI models and inference based on them.

Shaping a service is about understanding the collaboration patterns between business and AI organizations. These patterns directly impact funding needs and how an AI organization presents its services to potential internal customers to manage their expectations properly. It also guides the AI managers in clarifying which aspects of a service are pivotal, core, or more peripheral as a base for investment decisions.

AI organizations that want to run AI projects need data science know-how, but also data engineering capabilities to efficiently make use of existing data and AI translator capabilities to bridge the gap between business needs and concrete requirements what an AI model should deliver – our capabilities triumvirate. Understanding the cost drivers and the budget structure were additional aspects, so was pointing out the main challenge of project-focused funding: balancing project acquisition to prove the organization's relevance and ensure funding vs. available resources to execute the acquired projects.

When AI organizations want to run and maintain AI models, they need additional capabilities: support and application management, integration engineering, and reporting and visualization. They have to manage the AI models and monitoring the model performance as two additional AI-specific operations tasks. Finally, process definitions and target operating models help AI organizations establishing a structure and ensuring smooth, day-to-day operations.

This knowledge is what distinguishes an AI manager from a conventional IT manager or a technology- and algorithms-focused data scientist.

CHAPTER 6

AI and Data Management Architectures

The AI capability in an IT landscape is more than just the icing on the cake. It is a core capability for forward-thinking, data-driven, innovative companies. While not just the icing, it is also not the whole cake. An AI organization has to integrate its components into the overall corporate application landscape – and its own system landscape typically consists of more than just a few Jupyter notebooks. In other words: This chapter looks at the big architectural questions any AI organization faces that is more than just a one-man show.

The architectural challenges relate to three topics:

- Architecting AI environments for model training and ongoing execution and inference

- Integrating AI environments into the corporate data management architecture

- Understanding the impact of the public cloud on AI environments

Architecting AI Environments

When it comes to an organization's efficiency and its capabilities, tooling is a central enabling factor. The tooling impacts and determines the services an AI organization can deliver and how efficiently they can work.

We looked already at Jupyter notebooks in an earlier chapter. They help to create and optimize AI models. Here, we broaden the perspective and look at the other enabling systems an AI organization needs to work efficiently (Figure 6-1).

© Klaus Haller 2022
K. Haller, *Managing AI in the Enterprise*, https://doi.org/10.1007/978-1-4842-7824-6_6

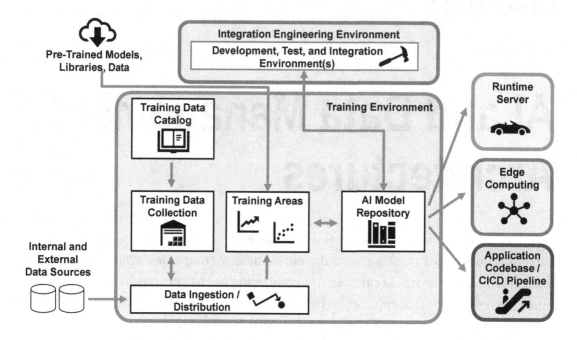

Figure 6-1. *Application Landscape for AI Organizations*

Ingestion Data into AI Environments

Training AI models and applying them to actual data for inference require adequate data. The data might come from operational systems, data warehouses, physical sensors, log files, or other data sources. The challenge is to build an architecture that delivers all potentially relevant data to the AI training environment or to the components performing AI inference in the production system using the trained AI models (Figure 6-2).

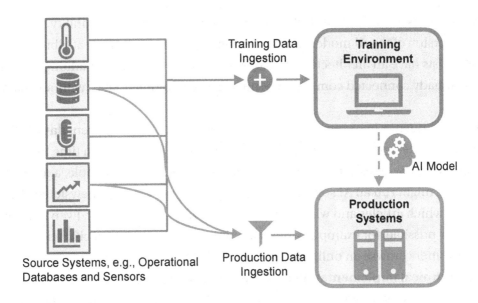

Figure 6-2. *Data Ingestion Scenarios for AI Organizations*

The training and the inference use cases are similar regarding setup and technical solutions, though two nuances differ. For **training data**, the rule of thumb is: **more is always better**. More attributes per data point increases the chance that more attributes turn out to be relevant for and contribute to the model. The model accuracy gets better, no matter whether the model predicts or classifies. More rows stabilize the model, and more tables allow for a broader range of use cases. If input variables are irrelevant, they simply do not become part of the final model.

Inference data serves as input for the trained, ready-to-use AI model. Such data describes the current situation for which the AI model makes predictions or performs classifications. For example, the data might describe the status of a chemical reactor, and the model predicts whether the reactor will blow up within the next hour. Such an AI model has a set of input parameters. Data scientists or automated tools must provide the current values precisely for all these input parameters. More data does not bring a benefit. Suppose the model needs the reactor's pressure five minutes ago. In that case, the prediction does not improve if we also know the pressure two days or one minute ago – or even tomorrow's weather forecast for the Sahara desert. The model cannot incorporate such uncalled for data into its predictions and classifications.

While not-asked-for data is useless for inference, architectural flexibility is advisable. The next version of the AI model after retraining might have two more attributes. Ideally, data scientists have an architecture that allows removing attributes or adding new ones from an already connected component with a reconfiguration, that is, without changing code.

A second nuance between training and inference data is the different **data delivery frequency**. Model retraining is an activity data scientists perform from time to time, for example, every few weeks. Inference takes place more often. For example, a marketing department might run an AI solution twice a week that personalizes newsletters by predicting which articles and which ads work best for which customer. Even real-time inference is possible, for example, when targeting customers with specific ads and items while customers browse an online shop.

The **technical implementation** for the data delivery or data ingestion can build on existing data integration patterns (Figure 6-3). Most companies run one or more suitable system already. The characteristics discussed in the next few paragraphs help to communicate the needs to internal data integration teams or external vendors and integration partners. Certainly, it helps as well if the AI organization has some understanding of the potentially possible patterns.

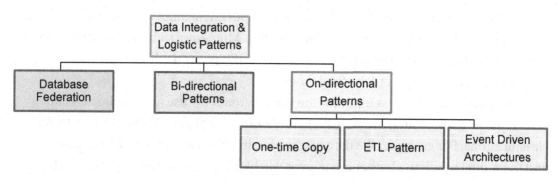

Figure 6-3. *Selected Data Integration Patterns*

The probably best-known database-related integration pattern is the **database federation pattern**. A federation defines how database schemata from various databases relate to each other. The pattern allows querying tables from multiple databases and even joining them without knowing and caring in which database they reside. When considering the two AI data ingestion use cases, the federation pattern does not help to get data from operational databases to training environments or production systems for inference.

Bi-directional patterns keep redundant data in two (or more) systems consistent. For example, a solution has one database in Zurich and one in Singapore to reduce data access latencies. Both store the same data. Bi-directional update patterns enable applications and users to write to their regional databases, not just one global master. Bi-directional patterns, such as "synchronization" or "correlation," propagate the changes done in one database to the other database and update the data there – entirely transparent for the users and applications.

Bi-directional patterns for training or inference data would promptly forward new or updated data from operational databases to the training and inference environment. However, they also have unwanted and potentially catastrophic consequences: Suppose a data scientist deletes the table with all payments of the current month in the training environment. Since every update is synchronized to the other database, the data disappears from the operational database. Not a good idea! Thus, bidirectional patterns do not fit the data ingestion use cases. We need a pattern that ensures that updates and data flow from the operational databases (or sensors, data warehouses, logs, etc.) to the AI environment, not vice versa.

Three **one-directional patterns** are of particular relevance. The first one is the **one-time copy pattern**. A database administrator or a data scientist copies data manually from the source systems to the training environment to train an AI model and run inference afterward. This pattern avoids investments into automating data copies and requires manual work. Typical use cases are a proof-of-concept, AI models supporting unique strategic decisions (expand to Latin America or the Pacific region?), or AI models requiring infrequent retraining with potentially changing source systems.

The second one-directional pattern is the **Extract, Transform, Load (ETL) process** pattern. It is well-known from data warehouses. The pattern has three steps: extract, transform, and load. They represent extracting the source system's relevant data, transforming the data to match the target schema, including data cleansing and consolidation, and loading the data into the target system. Data scientists or engineers implement these steps typically in SQL scripts. An orchestration component invokes and executes them, ensuring a stable and reliable execution, for example, every Sunday at 2:30 am.

The ETL process pattern is a typical example of batch processing. You collect data over time; then, the system processes all the collected data together at once. The pattern fits AI use cases that require data for periodical retraining or precalculating inference results. To give an example: A bank wants to know the product a client most likely buys additionally. At the beginning of each month, the bank updates the training data with the newest customer behavior and data and retrains its AI model. Then, the bank (pre-) calculates the next best product for each customer. When the customer logs in on his mobile app or contacts the call center, the system or the call center agent proposes this product to the client. It takes the information from the precalculated data, not requiring any online inference.

The ETL integration pattern is probably the most relevant for an AI organization supporting and optimizing business-critical operational processes and decisions. Calculating every week or month which customers might cancel phone plans is not the most innovative use case. Still, these are the ones ensuring the funding for many AI organizations since their business value is clearly measurable – and for such use cases, the ETL pattern works perfectly.

The third one-directional pattern is the **event-driven architecture** pattern. Changes in the physical or virtual world – a new sensor value or a customer clicking on a fashion item – trigger an event. Events flow through interconnected applications and components using technology such as Apache Kafka. The event routing is part of the business logic. It determines which systems see and process which events. There are two relevant differences to the ETL process pattern. First, the (near) real-time processing. Second, the publish-and-subscribe communication style. Components put events into channels such as "customer actions" or "weather sensor CDG." Other components subscribe to channels to receive and process these events. Potentially, they put the result into another channel to trigger follow-up processing. It is an utterly decentralized computing model.

Event-driven architectures allow for real-time processing. There is no benefit in delivering *training* data in (near) real-time since retraining does not occur so often. However, some use cases benefit from real-time *inference*.

The choice of the integration pattern and technology, however, should be well thought out. Switching the pattern or technology costs time and money. It is a complex undertaking for larger organizations. Thus, the AI organization should incorporate the life-cycle aspect as well. Which integration patterns are available in two or three years? What does the IT department plan for ETL tools? Event-driven architectures are seldom necessary, but using them can be a strategic move to protect investments since IT departments move in this direction. Ideally, the AI organization ensures it is aligned with the IT department's integration architecture.

So, to sum up, one-directional patterns are the solution for ingesting training data into training environments and for delivering data for inference to AI models. Choosing a fitting pattern for training and inference data ingestion is a child's play after answering three questions:

- How often do you build or retrain a model in your training environment?

- Is the AI model used for real-time inference, or are the inference activities performed as batch jobs, for example, at the beginning of each month?

- What is the integration architecture's roadmap for integration patterns and tools over the next years?

Storing Training Data

Training data ingestion patterns deliver data into the AI training environment, which has to store this data. A multitude of traditional and new technologies exist for storing data (Figure 6-4). AI organizations should make an informed decision on which ones to use in their training environment.

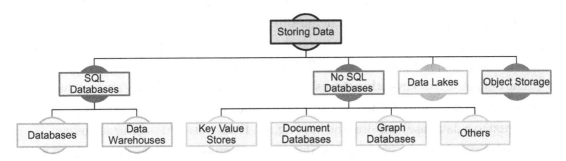

Figure 6-4. *Options for Storing Training Data*

The most popular form for organizing, processing, and storing structured, table-like data are **SQL databases,** under which we also subsume **data warehouse** technology.

Data warehouses are optimized for executing complex queries against large data sets most efficiently (OLAP – online analytical processing). Architects and database administrators rely on specialized schemata and table structures for more efficient query execution ("star" and "snowflake" schema). In contrast, OLTP – online transaction processing – optimized databases support higher transaction rates, that is, reading, writing, and updating one or a few rows. For most AI use cases, the difference between OLTP- or OLAP-optimized databases should not have a considerable impact (at least if there are not too complex queries for extracting data). Most probably, most data processing takes place in Jupyter notebooks, not in the database.

An SQL database for storing (some of) the training data is a must for an AI organization, though other technologies might be needed as well. The relevance of SQL databases bases on the fact that most (business and/or corporate) data reside in SQL databases in tables. Tables have the data structure AI training algorithms need as input. Thus, transforming data into a different structure to store the data usually causes costs without providing any benefits.

SQL databases have additional advantages. First, staffing projects is easy because SQL know-how is widely available on the market. Second, most IT departments have database administrators that can help with complex technical questions. The AI organization does not need specialized know-how. Finally, an AI organization needs just a plain-vanilla SQL database without advanced features. So, license-cost-free databases such as Maria DB are perfectly suitable.

While database and data warehouse technologies are from the last millennium, the rise of **No-SQL** – Not only SQL – **databases** are a phenomenon of the 2010s. The world of No-SQL databases is heterogeneous from focusing on scalability (by reducing on transactional guarantees) over schema-on-read (we see this with data lakes as well) up to different data structures. The latter is the focus here: key-value stores, document databases, and graph databases. They all store data differently.

The data model of **key-value stores** is truly minimal – pairs of a key and a value, such as <114.5.82.4, 20.07.2020 09:32>. Retrieving the value (e.g., the last time the IP address connected to a database) requires the key (e.g., an IP address). AI organizations can store such key-value pairs easily in SQL databases. Thereby, they prevent setting up a technology zoo.

Document databases store – surprise – documents. Documents are semi-structured, with the most popular formats being XML and JSON. Documents allow nesting attributes and do not enforce a fixed attribute structure. A typical JSON document looks as follows:

```
{
  "article": {
      "title": "Mobile Testing",
       "journal": "ACM SIGSOFT Software Engineering Notes"
      "author": {
          "firstname": "Klaus",
          "lastname": "Haller"
      }
  }
}
```

An AI organization should challenge the need for setting up one or more dedicated document databases in their AI training environment, even if source data comes from document databases. There are good alternatives, though SQL databases are usually none. Due to the semi-structured nature of documents, SQL databases with strict schemata are generally not a good fit. In contrast, data lakes and object stores are potential alternatives we discuss later in this section.

Graph databases make modeling and querying complex interconnected topics and relationships intuitive. Social networks are an excellent application area. A graph consists of nodes and of edges that connect nodes. Nodes can, for example, represent persons or topics. Both nodes and edges can have attributes to store additional information. Figure 6-5 contains various node types: individual persons, a band named Spice Girls, their albums, and some of their songs. The edges represent the relationships like being part of the band, having released a particular album, or a song being part of an album.

While applications based on graph databases can retrieve exciting results and insights, there is no technical requirement to perform such an analysis in a graph database rather than an SQL database. Graph databases cannot store data or data relations you cannot put into an SQL database. Graph databases are more an optimization. They ease writing queries for specific scenarios or allow for quicker query execution. Should an AI organization add a graph database to its technology stack? The answer depends on the circumstances. Does the AI organization have to run a graph database itself, or does the IT department have an expert team running them, or can they use a software-as-a-service graph database service freeing the AI organization from installation, maintenance, and patching? Is this data widely worked with, and does the graph database store important, relevant information? How is the intention to use the graph-structured data for training AI models? Are there performance issues in

SQL databases that cannot be solved otherwise? Graph databases require a business case to validate that they bring a financial benefit compared to not using them in the AI environment.

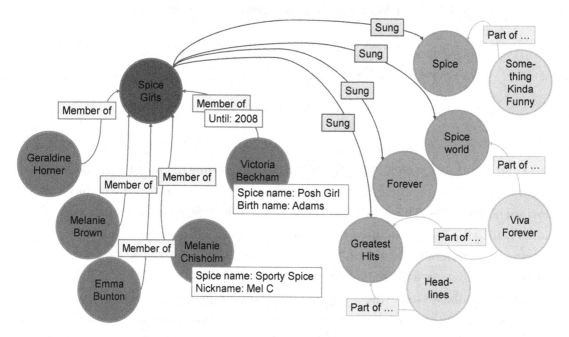

Figure 6-5. *Representing the Spice Girls in a Graph Database*

A **data lake** is a suitable choice for massive amounts of data stored in files, even petabytes. In contrast to a simple file system, data lakes allow searching for the content of files and aggregating information. For example, count the number of "Potential Breach" log entries in the folder structure 2021/07/* and its subfolders to understand whether there are more incidents than the month before. Data lakes implement a schema-on-read functionality but operate – in contrast to document databases incorporated into applications – on massive data sets covering data from various applications. There is no need to define a data structure or schema, just where and how to look for relevant attributes within a file. Thus, the data lake continues executing a query without throwing an exception if not finding the attribute, implying that there is no direct feedback if specific attributes do not exist – other than in case of SQL databases.

AI organizations earlier or later have to deal with semi- or unstructured data. They have to store such data somewhere, and, typically, SQL databases are not adequate. Is a Hadoop data lake the right choice? It requires effort to set up and maintain. If you are a big organization, maybe. Smaller organizations can use a data lake service in the cloud.

Finally, **object storage** is a storage option worth mentioning as well. Since the launch of AWS S3, the relevance of object storage is getting bigger and bigger. What was a filesystem in the past is object storage technology today: the place to put files and documents. It is the "standard storage type" in the cloud. In contrast to traditional files, manipulating files is not possible, just replacing them, which does not impact the work of an AI organization. However, it makes sense to invest some time to elaborate on how to deal with unstructured (and semi-structured) data.

Data Lakes vs. Data Warehouses

Data lakes are a popular choice for AI organizations to capture and store massive amounts of data, including text, audio, video, and sensor data from various source systems. At the same time, companies also run substantial data warehouses, which are especially strong for reporting and analyzing large amounts of (business) data. While data lakes and data warehouses might look like redundant concepts, they are not. AI organizations benefit from both, though technical processes and the commercial aspects differ.

What ETL – Extract, Transform, Load – is for data warehouses is ELT – Extract, Load, Transform – for data lakes. The steps are the same; their order differs (Figure 6-6). Deferring the transformation is a game changer cost-wise. The transformation step is the most time-consuming and expensive one. It covers cleansing plus ensuring consistency. Ensuring consistency is an analysis-intense, manual task, especially if several databases and reports contain similar, but slightly differently defined key performance indicators. Discussing with various experts and managers in different business teams and achieving a consensus can take several weeks. The benefits: A data warehouse has "gold standard" data quality on which engineers and business users can rely, and that is why managers fund the data warehouses. Thus, avoiding the cost-intensive transformation step before loading data into the data warehouse is no option. As a result, adding single attributes is already an investment, limiting the option to add data quickly. The management has (always) a say in whether and what data the data warehouse teams add.

In contrast, data lakes store uncleansed raw data. There is no discussion about single attributes; there is a decision on which database to add or which folders with massive amounts of log files. Adding data incurs no high costs, neither for analysis in the project nor for storage afterward. These low costs are the success factor for data lakes. Engineers can just add data based on a vague hope someone might have some kind of idea somewhere in the future to use the data. This someone, then, pays for the transformation, including cleaning and preparation.

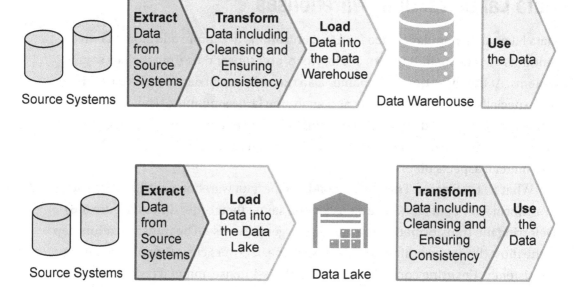

Figure 6-6. *Ingesting Data into Data Warehouses (Top) and Data Lakes (Bottom)*

Data Catalogs

A crucial prerequisite for high-quality AI models is adequate training data. Which customers are likely to terminate the contract in the following weeks? Which screws on the assembly belt have a defect? Data scientists can answer (nearly) all questions of the world if they have access to the necessary data for training their machine learning model.

Data warehouses provide large amounts of well-structured, well-documented consistent data. In contrast, data lakes collect a multitude of what data warehouses store, but without similar documentation – not what data scientists love, but an integral part of any data lake business case, as elaborated some paragraphs earlier. Plus, operational databases and other data storage means can contain additional data of interest for data scientists. A data catalog contains **information about data** in the various data sources, be it operational databases, data lakes, or other data. It helps AI projects **finding** potentially relevant **training data** they do not know yet. A data catalog makes the difference between a useful data lake and a useless data swamp. It is an enabler and speeds up the work of AI projects.

Data catalogs can provide information on the data attribute and the table level. Table level information consists of three elements (Figure 6-7). First, a data set has a name and a unique identifier. Second, a data set has a description of what this table is about, typically enhanced with category information based on a company's classification system and keywords that allow potential data users to find data quickly. Third, there is additional meta-data such as publisher, when it was published, by whom, data lineage info, etc.

Measurement results for radioactivity in plain yoghurt, 3.5% fat (ID: 54961665)	
Description:	*Issuing Date:* 22.9.2014
Measurement data for monitoring radioactivity in the environment, in food and feed	*Last Modified:* 27.1.2017
	Publisher: Institut für Hygiene und Umwelt, Hamburg
Categories: Energy, Environment	*Temporal coverage:* 4.2.2013
Key Words: Feed, Food, Radiation Protection, Environment	*Update interval:* n/a
	Regional coverage: Hamburg

Figure 6-7. *Data Set Description of a Publicly Available Data Set*

The data set description is the most crucial information in the data catalog. It enables data scientists to search for and identify potentially useful data sets that help training an AI model they work on currently.

Data catalogs also provide information on the attribute level, including the technical data type (String vs. Integer) and, ideally, the domain data type (e.g., whether a string stores employee names or a passport ID). The data catalog might even provide minimum and maximum values for each column. Figure 6-8 illustrates these two levels.

Domain data types of attributes and data set descriptions reflect one typlical data catalog use case: Finding all the columns with IBANs, customer name, or patient records, which is essential for legal and compliance teams (e.g., in the context of GDPR), but not really helpful for data scientists.

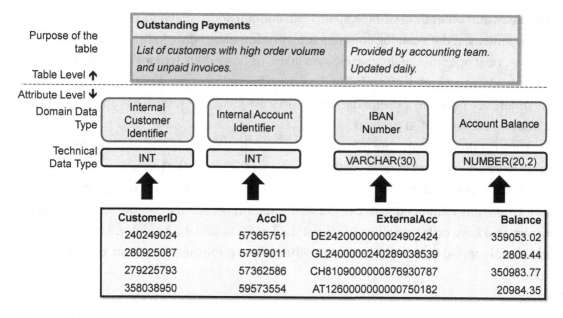

Figure 6-8. *Understanding Data Catalogs*

Data warehouses come with data catalogs and glossaries – and everybody knows that they are time-consuming to set up. So, can AI organizations or IT departments reduce costs and manual work by automating creating data catalogs data lakes with **data loss prevention (DLP) tools**, for example, with Symantec or Google Cloud DLP?

The short answer: **legal and compliance** teams benefit, data scientists not. DLP tools are great for finding domain data types. They identify which tables or files of databases and data lakes contain, for example, customer names, patient records, or religious affiliation information. Thus, they are perfect for identifying GDPR-relevant information. They do not help to understand the more sophisticated **semantics** of files and tables. Are these all customers with outstanding payments, or are these really important Swiss customers? Was the radiation measurement from yesterday or a week after the Chernobyl catastrophe? Collecting and compiling such information requires human work. Stringent enforcement of maintenance processes is the only way to ensure everybody does her tasks in this context. IT departments must technically prevent any upload of data to a data lake or the AI training data collection if the uploader does not provide a description with all information relevant for data scientists.

From Excel and MS Access to dedicated commercial software, AI organizations and IT departments have many options for data catalogs. Reasons for opting for sophisticated and more expensive solutions are their integration of crowd-intelligence features and their support for **governance processes** – the latter matters, especially in times of GDPR or with ethical discussions becoming more critical. Sophisticated processes and adapters to systems enable specialists to work more efficiently, though this is more a compliance than an AI topic. **Crowd-intelligence** mimics advice from senior experts who help new colleagues. A senior expert would direct her younger colleagues to data sets they need for their work. She knows where to find the data when the bank clients went to which branch. She knows where to find the clients that clicked on the details page for mortgage products in the mobile banking portal. Data catalog features mimicking such advice base their advice on statistics about frequently used data sets and attributes or typical queries submitted in the past. Such advice does not need direct human input. A less high-tech approach to crowdsource catalog information is to let data scientists (or other data users) rate the relevance and quality of data sets. The latter one, for example, is implemented in Qlik's Data Catalog using three maturity levels: bronze for raw data, silver for cleansed and somehow prepared data for data analysts to work with, and gold for reports directly useful for business users. Incorporating crowd-intelligence is still at an infant stage, though potentially soon a game-changer for the productivity of junior data scientists or data scientists new to a company.

To sum up: Data catalogs are crucial for benefitting from all data in a data lake and from less-known data sources in a company. They are a matter of discipline and stringent processes when establishing and maintaining the data catalog. They are a big help for compliance and governance issues as well – and the upcoming crowd-intelligence functionality is a great chance to increase data scientists' productivity.

Model and Code Repositories

Repositories are a must-have technical capability for AI organizations aiming for operational smoothness. We introduced them briefly earlier. They reduce the risk of having no backup copy of a crucial model or mixing up AI model variants and versions. Plus, they ease collaboration when multiple data scientists and engineers work on the same or similar models.

Repositories act as a hub for all AI models and additional data and store the following:

- The model's **purpose**, that is, a description of what the model achieves

- The **AI model** itself, for example, a Jupyter notebook file, including its previous versions

- **Other code** with relevance to production usage (interface code from integration engineering) or for reproducing the model (data preparation and cleaning scripts)

- **Experimentation history** documenting the model quality during the training phase, including architectures and (hyper)parameter settings

- **Approvals** for production usage, for example, a workflow action like an "approval click" or an email uploaded to the tool

Repositories for AI models can have various levels of sophistication, from just using GitHub to integrated MLOps platforms from public cloud providers such as the Microsoft Azure Machine Learning MLOps or specialized AI vendors such as Verta AI with fancy dashboards and CI/CD deployment pipeline integration. Since models are business-critical intellectual property, securing and protecting the repositories is mandatory.

Executing AI Models

Superior AI models are great for improving a company's operational efficiency if integrated into the organization's operational processes. In other words, applications invoke AI models for predictions or classifications. Besides precalculation without any integration, we discussed two approaches for technical integration earlier: **AI runtime server** and **integrating models into the application's code-base** by reimplementing them. We already discussed these topics from an integration engineering and a testing perspective – here, we focus on more architectural aspects.

Integrating the model by reimplementing them as part of the application code means that the architectural responsibility is not within the AI organization. The software architect and the software development team are responsible. The same applies to running their software solution.

There are various implementation options for an AI model runtime environment, mainly AI platforms from commercial vendors, open-source solutions, and the unique setup of edge computing/edge intelligence.

Commercial AI Vendor Platforms such as SAS or Dataiku or public cloud providers have a common and straightforward sales proposition: convenience, user-friendliness, and high productivity. They provide AI runtime environments easing not only the model training but also deployment and usage. Customers pay for these convenience and productivity benefits. Plus, they implicitly accept a vendor-lock-in, for example, if data scientists cannot transfer the data preparation and cleansing on a new platform.

One important remark. There are de-facto commercial platforms looking like "free" or marketed with providing "open source" components. They might even not charge any fees for software or integrated development environments and marketing using open-source components and industry standards. Still, they can generate a (cloud) vendor lock-in. You might not pay software license fees, but you pay by not being able to shift your workload quickly to other (cloud) vendors. When opting for "free" or "open-source" technologies in the cloud, AI organizations should check this detail because accepting a vendor lock-in opens up many more opportunities when choosing an AI platform.

Architecting an **open source-based AI platform** means putting together various open-source components to build an environment tailored to the needs of a concrete AI organization. For example, training models in Jupyter notebooks, using GitHub as a repository and pushing the results packaged as a Docker container to a public cloud provider for scalable execution or to run them on an internal cluster. The AI organization (and their internal customers) can discuss who runs and monitors the models or whether they become part of the actual code-base.

Finally, there is **edge intelligence**. Edge intelligence means that inference does not occur only in the company's data center or a cloud data center, but that there are edge servers deployed in all regions taking over the AI inference (Figure 6-9). The physical proximity solves latency issues, for example, when devices in the Australian outback invoke a prediction service on a server in Iceland.

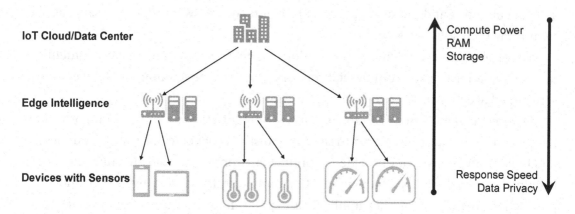

Figure 6-9. *Edge Intelligence*

Edge intelligence makes sense for IoT solutions with devices with limited compute and storage capacity. The assumption is that these devices cannot run AI interference locally. Over time, probably a second use case will gain more importance: protecting your AI models as crucial intellectual property. Companies might avoid deploying AI models to IoT devices to prevent competitors from getting access to a model by buying or stealing a device running AI models on them.

AI organizations can deploy an edge intelligence pattern themselves. However, relying on a cloud provider eases the deployment, primarily when potentially benefitting from other features, for example, integrating IoT devices and using the AI or data management features. To prevent from getting too naïvely into a cloud provider lock-in situation, having a transparent AI and cloud strategy helps – for IoT and edge intelligence and your AI platform.

AI and Data Management Architectures

While the previous section was about AI environments, we look now at how the AI environment fits in and benefits from an organization's overall data management architecture – and how AI relates to similar other services. Figure 6-10 provides a high-level overview with all relevant components and concepts, including, obviously, data warehouses.

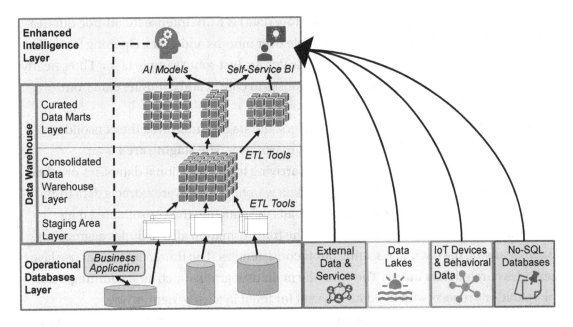

Figure 6-10. Architectural Blueprint for Traditional and New Business Intelligence Architectures

AI and Classic Data Warehouse Architectures

Databases and data warehouses are not only helping to store data in AI environments. Companies use them already for decades. A pre-AI architecture for data-driven companies consists of an operational database layer and a data warehouse layer.

Operational databases serve business applications by storing and processing data. For example, they store master data, product descriptions, bank transfers, or point of sales records.

Business users, potentially with the support of database administrators, can submit ad-hoc queries against the data. Did our marketing campaign targeting golden-agers result in more customers from this age category signing new contracts last month? It is an example of an ad-hoc query with a clear analytics focus.

Querying operational databases comes with two **limitations**. First, there is a potential impact on operations. Long-running complex analytical queries can impact the database stability and performance and, thereby, the overall system. It might not be suitable to submit such a query against the database of an online shop. The second limitation is the focus of operational databases. They (typically) store only data of one application or one specific domain. Combining data from different databases (e.g., one

167

with online shop data and one with retail store data) is work-intense for ad-hoc queries and requires high efforts to keep them stable over months and years. Opening firewalls or dealing with service accounts and certificates are just some sample tasks. Thus, nearly all companies run one or more data warehouses to collect and combine data from various operational databases.

A typical **data warehouse** has three layers: the staging area layer, the consolidated data warehouse layer, and the curated data marts layer. The **staging area** is an intermediate storage place for copied data arriving from operational databases or other data sources during the ETL process. The data warehouse software extracts data from operational warehouses, transforms, and loads the data into the consolidated data warehouse layer during the ETL process. The transformation also includes aggregating data, cleansing the data, and sorting out inconsistencies. The third data warehouse layer provides curated **data marts**. They provide main user groups such as accounting, sales, or operations reports with the data relevant for them in a user-friendly way.

From an **AI organization's perspective**, a data warehouse with its data marts and the tables in the consolidated layer is a big opportunity. It is a source of high-quality training data with company-wide agreed-upon semantics and from all areas of a company. Building a similar solution is often a multi-million, multi-year endeavor any AI organization should avoid. However, data warehouses have two clear limitations. The data is not real-time data. ETL processes typically run overnight to minimize the impact on operational databases during office hours. Thus, data is at least one day old, but companies might not refresh certain data more than once per month. Second, data warehouses are inflexible when it comes to adding new data. The rest of the company expects the data warehouse to deliver consistent data, requiring an in-depth analysis before adding an attribute – or one or multiple tables. Still, data warehouses are an excellent chance for AI organizations to kickstart activities in new areas with some initial, high quality, and "investment-free" training data.

In such an ideal scenario, the AI organization builds its solution as a new top layer, basing on the data marts or the consolidated data warehouse layer, as illustrated in Figure 6-10. The less the AI organization deals directly with operational databases, the more cost-efficient and agile it can operate.

Self-Service Business Intelligence

Naming self-service business intelligence (SSBI) tools the biggest fraud in recent data management and analytics history might be a slight overstatement. The problem is not their functionality, but what they are sold for to customers. Many vendors try to make their customers believe that the tool enables them to perform similar analytics as data scientists and AI algorithms can do – and this is not correct.

As the architecture in Figure 6-10 illustrates, SSBI tools are on top of data warehouses or data marts (as AI components are as well). The tools **enable business users** to self-define and create tables and views. In contrast to data warehouse views and tables, business users might have a very narrow perspective, not looking left or right or whether a similar performance indicator already exists. Speed, convenience, no effort are the usual foci of business users. Besides all criticism, SSBI tools solve a business need and ease the collaboration in corporate functions such as controlling in worldwide organizations. When a job is about copying data together from various Excels and cleaning and consolidating this information, SSBI tools can greatly help due to their collaboration features. However, when AI organizations consider using data from SSBI tools, they should be aware that the data quality might be less reliable than data from a data warehouse or an operational database.

SSBI tools become an issue for an AI organization and the company's innovation potential if it **hinders the adaption of AI**. SSBI tools are gold-plated Excel alternatives with more features to play around with, but not enabling business users to manually generate insights and models on the same level with the result of AI projects (and providing AI algorithms without explanation might also not result in useful models). The issue with SSBI tools is not their features, but if business users insist on "analyzing themselves" AI problems using SSBI tools instead of letting a statistical or machine learning algorithm do that much better or trying to create AI models without understanding the needed methodologies. This is a big risk or disadvantage with SSBI tools, and it might cause issues with the standing of an AI organization within the company. AI managers should beware.

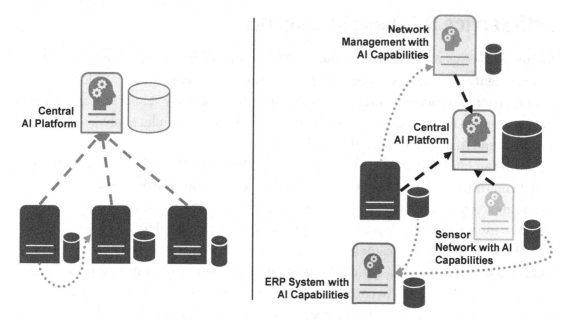

Figure 6-11. *Classic AI Architecture with an AI Platform Being the Only "Intelligent" Solution (Left), Pantheistic Intelligence in Today's Landscapes with Various Components with AI Capabilities Incorporated (Right)*

Pantheistic Intelligence

Pantheistic intelligence reflects that neither data warehousing teams nor AI organizations have a monopoly on collecting massive amounts of data and generating additional insights with AI methods. Software **vendors enhance conventional products** with AI features, whether they develop core banking systems, building control applications, or network management and monitoring solutions (Figure 6-11).

For example, a network team might want to speed up the resolution time for customer incidents by identifying a root cause quicker. A typical scenario: a WLAN router crashes, causing hundreds of devices being not reachable anymore, and resulting in hundreds of incidents from monitoring. This flood of incidents hinders network support engineers from identifying the crashed router as the root cause.

In such a situation, the AI organization could offer help to train a model for identifying root causes when there are several "real" incidents causing massive amounts of secondary, non-root-cause incidents. AI specialists train a model based on very few cases. They get the needed training data from a monitoring team, or by creating interfaces to all network and hardware devices. But there is a quicker solution: The network management buys a network monitoring and management solution that comes with interfaces for devices from all important manufacturers and contains an AI component besides the classic network management and monitoring features. Moogsoft is such a software provider. You have to provide a lot of data and do a lot of training, but still, would you go for the ready-to-use standard software or choose self-development if you are the head of networking in your organization?

Pantheistic intelligence is a reality. More and more software vendors incorporate AI functionality in their products. Thus, AI organizations might gravitate to strategic applications with a direct customer and business impact, for which the company wants to integrate data from various sources to build a superior model to achieve a competitive advantage.

New Data Categories

Master data and transactional data, this was what most systems could store some years ago. Today, new data categories are standard such as log and behavioral data, IoT data, or external data.

It was a long time ago when online fashion shops just kept your shipping address and order history. Companies today are data-hungry and store and analyze **behavioral data**. How do customers navigate in an online fashion shop? Which items do they scroll over, and which things make them pause for one – or ten – seconds? Often, such data comes from **log files**. **IoT devices** deliver sensor data such as pressure and temperature information – or pictures and video streams. They are core building blocks for digitizing new business sectors such as logistics, manufacturing, or production – or the basis for innovating the core products of technology and industry companies.

If companies do not have the needed data internally, they rely more and more on **external data** providers. External data can be "paid for" data, for example, regarding customer creditworthiness or validating shipping addresses. Many countries and public offices also make their data publicly available for free, for example, as part of "open data" initiatives. Such external data often comes in an easy-to-use form, such as tables in CSV files. When AI organizations identify such valuable external data to be beneficial, they can easily integrate the data. In contrast, behavior data, log file data, and IoT data typically require more data preparation until an AI organization can integrate the data and use it for training purposes. Such investment might not pay off for each small AI case. Still, it is an excellent chance for optimizing AI models that have a considerable impact on corporate success, be it customer understanding or optimized logistics.

Cloud Services and AI Architecture

The public cloud providers open a gigantic innovation potential, especially for SMEs, to integrate highly innovative, ready-to-use technologies without upfront investment. They change how IT organizations work, how they think and approach AI, and are the most significant influence factor on AI architectures for the 2020s.

Cloud computing builds on three main pillars:

- Infrastructure as a Service (IaaS)

- Platform as a Service (PaaS)

- Software as a Service (SaaS)

Best known are IaaS and SaaS. Everyone has been using **Software-as-a-Service** solutions already for years: Google Docs, Salesforce, and Hotmail are well-known examples. They are convenient for users – and the IT department. SaaS takes away the burden of installing, maintaining, and upgrading software. Using an AI solution in the web, for example, managed Jupyter notebook environments, makes own installations obsolete – though the effort for integrating them with other company systems should not be underestimated.

IaaS – **Infrastructure-as-a-Service** – thrives as well. Many IT departments migrated some or all their servers to the cloud. They rent computing and storage capacity from a cloud vendor in the cloud provider's data center. Thus, IT departments do not have to buy, install, and run servers anymore – or care about data center buildings and electrical installations. IaaS brings major benefits for AI organizations because they solve the issue with highly fluctuating processing needs. AI organizations do not have to buy hardware with enough capacity for extreme loads for training large AI models once a month. Instead, they rent as much as necessary and when necessary. They get immediate and unlimited scalability, high reliability, or any combination of that.

IaaS and SaaS revolutionize IT service delivery. Pushing a button gets you a VM. Office O365 makes installing software patches on company servers obsolete. Quicker and cheaper – but neither IaaS nor SaaS enables you to build revolutionary new services and products for your customer. PaaS – **Platform-as-a-Service** – opens up this opportunity. PaaS is a game-changer for fast-paced innovations. Every AI department should closely monitor the innovation pipeline of the cloud providers.

With PaaS, software developers and AI specialists assemble and run applications and services with ready-to-use building blocks such as databases, data pipes, or tools for development, integration, and deployment without dealing with complex installations or IT operations. The real game-changer is the massive investment of the big cloud service providers into production-ready AI services from computer vision to text extraction, aiming to make state-of-the-art AI technology usable by developers without AI background. Plus, they have exotic-obscure offerings such as ground station for satellite data or for building augmented reality applications – and they allow third party providers to make their software available, hoping for a network effect as known from the iStore or the PlayStore.

These PaaS services also impact AI organizations, and not every data scientist might be happy about every change. If a cloud provider offers image recognition, there is no need and opportunity to train generic AI models anymore themselves. Application developers use these services instead of asking the AI organization to train an AI model. For example, they need as much time to integrate AWS's Rekognition service as for writing a printf-command – and as a result, they know whether a person smiles on a picture and whether there are cars on the same image. Commodity AI services will come from the cloud. Consequently, AI organizations might have to move to more complex, company-specific tasks and build and maintain more extensive AI solutions – or lend out engineers with knowledge of, for example, Azure's AI functionality.

On the other hand, AI services from the public cloud require some governance. IT departments and AI organizations should not be too naïve. First, cloud services are not free. Business cases are still needed, especially if you are in a low-margin market and need to call many expensive cloud services. Second, the market power changes. Some companies pressed their smaller IT suppliers for discounts every year. Cloud vendors play in a different league. Third, using specific niche services – the exotic ones which help you to design unique products and services to beat your competitors – result in a cloud vendor lock-in.

The cloud-vendor lock-in for platform-as-a-service cannot be avoided. Enterprise, solutions, and AI architects must manage this unpleasant reality. A simple measure is separating solutions and components based on their expected lifetime. "Boring" backend components run for decades. IT departments must be able to move them easily to a different cloud vendor with little effort. Then, there are innovative, short-lived services, mobile apps, or web-frontends. They have a life expectancy of less than a decade – plus, every new head of marketing or brand officer demands fundamental changes. For such solutions, vendor lock-in is of less concern. You can change the technology platform anyhow every few years when developing the next version.

The implications for an AI organization are crystal clear:

- Data preparation and data cleaning activities are probably the artifacts that have the most extended lifespan. Being able to reuse them when changing the platform is essential.

- The platform for "do-it-yourself" AI model development does not matter too much. All platforms support linear regression and neural networks. Retraining these models is less of an issue if the data preparation works.

- PaaS offerings such as labeling support, object detection, or speech recognition result in a high vendor locking. They are not standardized; functionality and interfaces differ.

Vendor lock-ins are painful. Some vendors are feared for their creativity to play around with licensing and pricing every year. On the other side, organizations cannot prevent any form of lock-in, especially if the alternative is missing out on the quick wins of AI PaaS services. Being attacked by more innovative competitors using the power of PaaS is not a viable alternative. IT departments might ignore cost-savings and agility improvements IaaS and SaaS promise, but AI organizations should be careful when rejecting AI PaaS services as innovation opportunities.

Summary

The first architectural challenge for AI organizations is to organize their AI environment for training AI models and, potentially, for inference using the trained AI models. There are apparent components such as training areas with Jupyter notebooks, plus data storage such as databases or data lakes for the training data. However, only well-maintained data catalogs actually lift the data treasures in data lakes – and without data ingestion and data integration solutions in place, updating the training data is a nightmare. Model management helps keep track of the existing AI models and their versions, preventing chaos such as losing models or deploying outdated ones. AI organizations might also manage AI runtime environments for inference – or solutions teams take over the models, integrate them in their code, and run the application without the involvement of the AI organization.

AI organizations can only efficiently operate if they get their hands on the necessary data. Thus, it is vital to be aware of the options such as directly exporting data from operation databases, IoT devices, data lakes, or external sources. Still, the best sources are data warehouses: well-documented, well-maintained, consistent data covering many domains. A data warehouse enables AI organizations to create AI models for various new areas or business questions quickly.

In the coming years, more and more AI environments will move to the cloud. Paying only for compute resources when training a model is a significant booster for resource effectiveness. However, the biggest change to AI organizations is the AI PaaS services making training standard services such as plain-vanilla image detection obsolete and freeing up data scientists to work on the tough company-specific topics.

CHAPTER 7

Securing and Protecting AI Environments

Once upon a time, IT and information security were simple. IT security specialists deployed malware agents and set up a firewall – and the organization was safe. Today, information security is much more exciting and complex, requiring everyone's contribution. Even data scientists in AI organizations must be aware and contribute to security – and AI managers might suddenly get accountable and responsible for proper security measures for their organizations' AI infrastructure. The pressure is ubiquitous, but not the necessary know-how – especially not in AI organizations. To complicate matters, many senior IT security and risk professionals do not have an in-depth understanding of AI methodologies and techniques they assess. A dilemma for which this chapter provides the solution by addressing four topics:

- The CIA triangle describing the central aims for IT and information security

- Collaboration patterns and responsibility distribution between IT security and the AI organization

- AI-specific security threats and risk scenarios, including potential attackers and assets under attack

- Concrete technical and procedural measures to mitigate AI-related risks and improve the overall level of security

Figure 7-1 visualizes their interdependencies. The following pages elaborate the topics in detail.

© Klaus Haller 2022
K. Haller, *Managing AI in the Enterprise*, https://doi.org/10.1007/978-1-4842-7824-6_7

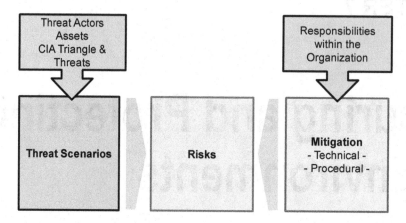

Figure 7-1. *IT Security Risks – The Big Picture*

The CIA Triangle

Information security professionals frequently refer to two fundamental principles of information security: the need-to-know principle and the CIA triangle or triad. CIA is an abbreviation for confidentiality, integrity, and availability (Figure 7-2). These three terms make it clear: information security is more than protecting your network with firewalls and running anti-virus software. Thus, they apply to and impact AI organizations and their infrastructure and platforms, too.

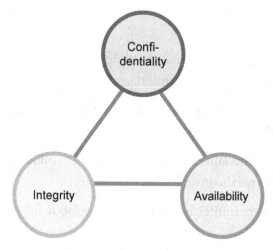

Figure 7-2. *The CIA Triangle*

Confidentiality means keeping secrets secret. Users can only access data and systems they need for their work (need-to-know principle). Typically, employees get necessary access based on their roles. Other access rights require someone to grant access explicitly. A data scientist working for HR needs access to HR-related data stored in the data lake. In contrast, a data scientist working on sales topics does not have access to HR data.

The second aspect of the CIA triangle is **integrity**. Integrity means that data and information in databases, files, and data lakes are correct. Applications, users, and customers can rely on them. For example, training data is highly critical for creating a usable AI model. Removing all data from a country or related to particular minorities impacts the correctness or quality of the model. Integrity does not require making all data manipulations and tampering technically impossible. Ensuring integrity requires detecting manipulations, for example, by using and validating checksums. Thus, integrity relates closely to non-repudiation. A user or engineer can only change data, applications, or configurations after authenticating herself. Authentication is a prerequisite for logging who did what. Needless to say: It is essential to store these log files tamper-proof.

The third and final concept of the CIA triangle is **availability**. Data and information stored in files, databases, and software solutions deliver only benefits if one can access the data. When a building access system relies on an AI component to identify employees, the system must be operational 7/24. If the system is not available, nobody can enter the building. Typical technical measures for availability are backups or redundancy. The architecture must accompany these measures with additional protection against denial of service attacks or unintentional application shutdowns.

Security-Related Responsibilities

For technical and AI-focused specialists, security is often just a sideshow. Most engineers and architects enjoy putting the various AI-related components and systems together. When projects face the issue of who takes over a particular task, there is usually a solution within the project team, even if there is a shortage of engineering capacity. The challenge is more to keep the systems secure for the years they run. Who periodically inspects logs for irregularities? Who checks whether vulnerabilities emerge and vendors provide patches – especially for zero-day attack vulnerabilities?

When AI organizations run training environments themselves, they are responsible for runtime servers and might have model repositories or data lakes. The AI organization must ensure that these systems are secure. It is often unclear which tasks a security organization takes over and which ones the AI teams have to organize themselves. A particular challenge for AI managers is understanding a typical double role of IT security organizations: security services and security governance (Figure 7-3). Both help a company and an AI organization to secure their IT infrastructure and their application landscape. Approach and aims, however, differ.

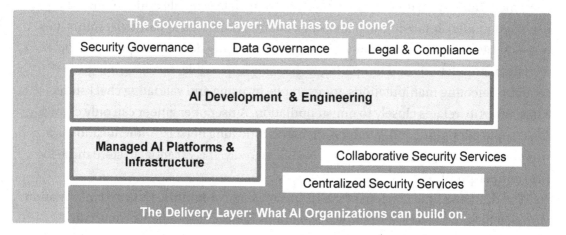

Figure 7-3. *An Organizational Perspective on Security Responsibilities*

The company's **security governance** function's core tasks are elaborating policies, performing assessments, and managing IT-related operational risk emerging from derivations of actual systems implementations and procedures from policies. The security governance function is just one governance function. There are others such as legal and compliance or data governance impacting AI organizations as well, which other chapters address.

Writing security policies is like writing a wish list for Santa Clause. You write them down; the actual delivery and implementation are not your concern. Policies contain procedural and technical requirements. One example of a procedural requirement is demanding penetration tests before software developers or data scientists make a web service available for the rest of the company. An example of a technical requirement is demanding to encrypt all network traffic with TLS 1.3. Governance specialists elaborate policies often from standards and frameworks such as the ISO 27000 family and ISAE 3402 or best practice documents from the software vendors or public cloud providers.

The IT security governance teams do not provide any technical services to other teams. They guide the organization to a more secure setup with their policies. Plus, they support senior management to understand the company's overall IT security and risk situation. The rest of the organization, including all IT and AI teams, act as Santa Clause. They have to fulfill the wishes and needs laid out in policies. Each **application owner** must ensure that his application is secure and complies with the security policies. AI organizations must fulfill the same requirements for their training environments, runtime servers, or model repositories as any other application or system handling large amounts of sensitive data. These tasks are usually with the AI organization's application management and operations specialists providing internal services for the data scientists, data engineers, and AI translators. Figure 7-3 refers to these components as **managed AI platforms and infrastructure**.

Besides the governance function, many security organizations provide security services – application owners inside and outside the AI organization benefit from them. One service type comprises **centralized security services**. These services address and solve specific, generic tasks for the company. They relate to protecting the company's perimeter – firewalls, data loss prevention tools, scanning incoming files for malware – or checking for patching-related issues or (OS-level) vulnerabilities.

In contrast, **collaborative security services** require close collaboration between an IT security delivery team and application owners. Identity and access management (IAM) or web application firewalls (WAF) are examples. Both require per-application customization. An IAM system can only manage access to an AI platform if it knows the roles and if the AI platform can interact technically with the IAM solution. Sophisticated WAFs need training for fine-tuning their rules to learn all the legitimate requests an AI runtime server might receive.

In other words, in contrast to centralized security services, collaborative security services require the AI organization to perform dedicated tasks to secure the AI systems. The integration engineers, who set up the managed AI platforms and infrastructure, are responsible for securing them. After the go-live, the application management and operations experts of the AI organization have to keep their systems secure for the years to come. The AI management can ensure that data scientists and engineers are not distracted by security-related tasks to keep up the AI organization's productivity. There is just one big exception: personal unmanaged installations and applications. When AI specialists install systems, they have to secure them, whether the systems run on servers or in the cloud.

Experimental installations pose a particular risk, though there can be a need for them. From time to time, data scientists or engineers try out new tools. They install the tools on their laptops or a VM and perform some use cases, potentially with actual company data. A typical misconception is that such systems are not relevant to the organization's overall security posture because they are just experimental. The contrary is true, especially (but not only) when they contain actual data. Experimental installations are often not secured as well as production systems. Thus, hackers might take over such experimental systems more quickly and misuse them for attacking other applications and extracting confidential data. If protecting such systems seems not feasible, the only solution is setting up a sandbox environment in a locked-up network zone.

AI specialists and managers must never forget two reasons why to secure their systems. The first one is to ensure that their data remains confidential and to guarantee data integrity and their systems' availability. They are responsible, even though they can or have to build on some security services.

The second reason why AI organizations must harden their systems is to prevent **"spilling over"** effects. A successful attack, for example, on a single (web) application, must never put the attacker in the position to take over effortlessly also the AI runtime server or the AI training environment – or experimental systems. No security organization can guarantee that. It is the ultimate responsibility of every member of the AI organization.

We conclude the discussion about responsibilities with a final remark about AI organizations using **public cloud services**. It is a booming topic, which means that governance and responsibilities are less elaborated than for traditional data centers. The AI organization should clarify with the security organization whether they are allowed to use the intended cloud services. Furthermore, they should also clarify whether the security organization provides the standard security services for the envisioned cloud environment. Otherwise, the AI organization might have to set up and manage the cloud security all by themselves – or risk that auditors shut down their environment.

Mapping the Risk Landscape

The rise of AI and its contribution to business-critical processes transform organizations – and creates new risks. What if attackers manipulate AI features or steal data or models? AI is a new domain for many risk and security specialists, but the methodology remains the same. First, understand the threat actors and their motivations, aims, and capabilities. Second, identify assets attackers might attack or try to steal. Third, understand scenarios and what attackers aim for. Fourth, and finally, filter for likelihood and potential impact to identify the relevant risks. Figure 7-4 illustrates these steps. We take a closer look at them in the following.

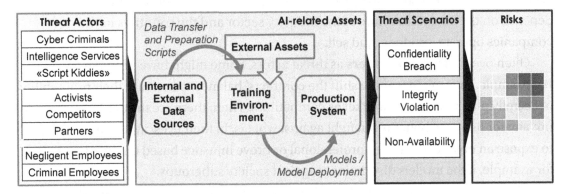

Figure 7-4. *From Threats to Risks for AI Organizations*

Threat Actors

Script kiddies, professional cybercriminals, and foreign intelligence services: a diverse and colorful crowd threatens today's IT systems and AI environments. Threat actors aim for stealing intellectual property, blackmail for ransom, or try to crash crucial servers. They threaten AI organizations, AI-related assets, and AI-driven business processes and control systems. The threat actors differ regarding motivations and capabilities, technical and financial-wise.

The hurdle for engaging in cyberattacks today is lower than ever. Attack tools are widely available, enabling **script kiddies** to pursue not-so-much-recommended leisure activities. **Cybercriminals** have more personnel and financial resources for attacks. They outsource tasks to criminal service providers when they miss specific know-how or run large-scale attacks. Cybercriminals have clear monetary motives, such as extracting ransom payments. For example, they encrypt all company data, including backups – and decrypt the data only after payment. They are a challenge for small and medium enterprises, but even for well-funded IT security teams. Even worse are **state-sponsored attackers** and intelligence services. They have nearly unlimited resources. Their focus is mainly on critical infrastructure, enabling them to bring down another country's infrastructure. Alternatively, they try to steal business and military secrets from governmental offices, the military, or highly innovative companies and institutions of strategic interest. AI models in antiaircraft systems might be of high interest to rivaling countries. So are the newest insights about autonomous cars or biotech innovations.

There are three more types of external attackers: activists, partners, and competitors. **Competitors** are natural threat actors since they benefit directly from illegal activities such as stealing intellectual property or harming ongoing operations. However, it depends on the business culture of the industry sector and the countries in which companies operate, produce, and sell.

Often overlooked are **partners** as threat actors. Some might have a hidden agenda. For example, they might want to shift the contracts' balance to increase their profitability or secretly plan to enter the market as competitors. Then, there are the **activists**. They do not aim for financial benefits but fight against what is, in their eyes, injustice. They want to expose an organization as unprofessional or prove injustice based on the AI models, for example, if the models discriminate against society subgroups.

When organizations mistreat their employees or employees try to make quick money, some turn against their employers and harm them. **Criminal employees** with access to significant data sources (e.g., data lakes) are particularly dangerous. Without logging, forensic experts cannot understand what happened or pursue criminal investigations. If everyone can copy large amounts of data, for example, to personal USB sticks, this further eases illegal activities.

Besides the risk of criminal attacks, there is the risk that good-willing **employees** make mistakes due to negligence. They might not understand security procedures; the procedures might be inconvenient or – from their perspective – not make sense. Such situations can result in employees following security procedures halfheartedly, potentially resulting in vulnerabilities.

Which threat actors are relevant for a company? More or less, every company is a potential target for script kiddies and cybercriminals. For example, cybercriminals often demand "only" five-digit ransom payments. Small and medium enterprises can pay such sums – and limited IT defense capabilities make them viable targets, especially if attackers can automate attacks. The risk of the other threat actors has to be looked at for each company individually. For example, a French corporation might come under attack from foreign activists, for example, due to tensions between France and Turkey.

After understanding potential and relevant threat actors, the next step is to look at what the attackers might target. In other words: What are the assets of an AI organization?

Assets in AI Organizations

"Asset" is a widely used term. It refers to software, hardware, and data – everything needed for information processing. Four asset types are essential for AI organizations:

- **AI models.** For example, they help optimize business processes, oversee assembly lines, control chemical reactors, or predict stock market prices. AI models can be an integral part of the application's codebase or run on AI runtime servers.

- **Training data** is essential for data scientists. They need the data to create and validate AI models.

- **AI runtime servers** are the software and hardware systems on which the AI model runs and inference takes place. It can be an RStudio Server, the model can be part of the actual application code, or the application implements an edge intelligence pattern.

- The **training areas** on which data scientists create, optimize, and validate the AI models, for example, using Jupyter notebooks.

Various additional systems and assets are of relevance for an AI organization that aims to work efficiently. They primarily relate to the training environment:

- Internal **upstream systems** and their data such as SAP, a core banking system, or Salesforce.

- **Pretrained models, libraries**, and reference **data** from the Internet, speeding up the training process or helping to get better models.

- The **data ingestion and distribution** solution, for example, an ETL tool, loading data from the data sources into the training data collection and, potentially, loading it as well into the training areas.

- The **AI repository** storing the model descriptions and training history of the models – and the models themselves.

- A **data catalog** indexing and describing the data available for training models.

Any AI-related threat or attack relates to one or more of these assets (Figure 7-5). When successful, attacks compromise one (or more) information security CIA triangle properties: confidentiality, integrity, availability. Some assets might impact production, others are "only" needed for efficient training. Understanding the details is part of the threat analysis.

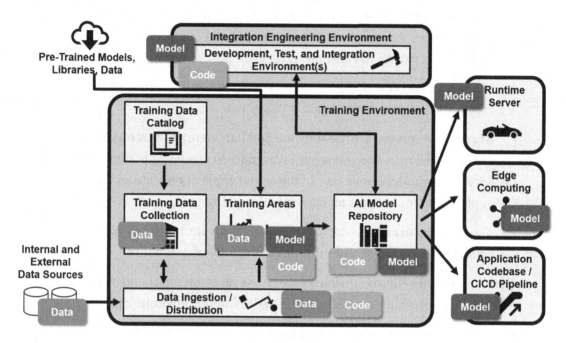

Figure 7-5. *The AI Ecosystem with Its Assets*

Confidentiality Threats

AI-related systems need and work with large amounts of data to generate new insights. These systems are of high interest for threat actors trying to **steal intellectual property**, particularly **training data** or **AI models**. They are a gold mine for competitors time-wise and money-wise. For example, Waymo, a self-driving car company, prides itself on more than 20 million self-driven miles and more than 15 billion simulated miles. When competitors get this training data or the AI model built with this data, they save millions of investments and years of work. Likewise, a stolen underwriting model in the insurance sector enables competitors to lure lucrative customers, while staying away from problematic ones.

Just by looking at Figure 7-5, it becomes obvious where training data resides. Attackers can steal it from the original **data sources** and databases. Still, the most extensive data collection is, in most cases, the AI organization's **training data collection**/data lake, comprising all available data. In contrast, the actual training areas, for example, a single Jupyter notebook, provide more limited data. Attacks on the data ingestion and distribution component are another way to collect data.

Besides stealing data, it is also an option to steal the final **AI models** to use them – or to understand how a company acts, for example, on the market and to undercut profitable offers. AI runtime server or other production systems, the AI model repository, and training areas store or have access to one or many of them. The AI **model repository** is the component storing all AI models, even with explanations and documentation. **AI runtime servers**, if used, contain an extensive collection of AI models. Other systems or components, such as the **training areas** or application source code – if AI models are reimplemented and part of the code – provide only one or a few AI models. However, for any insurance, the loss of even a single model for one large or important customer segment can be already catastrophic.

Loss of intellectual property might not be an issue for all companies. However, a second confidentiality threat is relevant for many companies: stealing data to extort a **ransom**. Hospitals store which customers are HIV positive. Insurance companies store salary-related data about their customer companies' employees if they handle sickness allowances. No hospital or insurance company wants to see such information on the web. Neither do their customers and patients. It is the perfect setting for blackmailing all of them if cybercriminals get their hands on such sensitive data.

In the past, **stealing files and data** required sophisticated attacks – at least for data from not-Internet-facing servers. Still, an AI organization would act foolishly in ignoring such risks, though there are more severe and more uncomplicated-to-perform attacks today.

With the advent of the public cloud, companies train AI models there. Training data and trained models are in the cloud. If the company or AI organization sets up the cloud security properly, assets are safe. However, wrong configurations happen quicker, and (weakly protected) Internet-facing public cloud environments are an invitation for attackers. Many companies had to learn it the hard way in the last years.

The most effortless and potentially most challenging to detect confidentiality attacks involve **employees**. In many organizations, they can transfer easily and risk-free any file – including models or training data – out of the company.

Stealing a model, however, is also possible without access to it, just by **probing**. Suppose an attacker wants to mimic a "VIP identification" service identifying stars and starlets on images. The first step is to crawl the web to collect sample images (Figure 7-6, left). At this point, it is not clear whether pictures show VIPs. So it is not possible to train an AI model with the data. Thus, the second step is to submit these pictures to the "VIP identification" web service. This service returns which images contain which VIPs (middle). Now, the attacker can build a training set for training his own neural network mimicking the original one with this information gained from probing and use it in production (right).

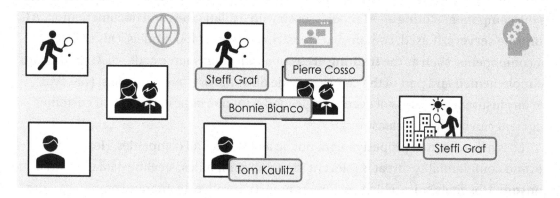

Figure 7-6. *Probing a Model for Recreating It – Trawl the Web for Sample Pictures (Left), Let the To-Be-Mimicked-Service Annotate the Images (Middle), Train a Neural Network Based on the Newly Annotated Images (Right)*

A final remark about non-financially motivated threat actors: **activists** can be interested in AI models. Their motivation would be to validate whether a model discriminates subgroups of the society systematically, for example, by pricing them differently than mainstream customers.

Integrity Threats

AI-related integrity threats relate, first of all, to the **model quality**. When threat actors modify a model or interfere with its training, the model classifies images incorrectly or makes unprecise predictions. Consequently, a factory might produce unusable parts, an aircraft carrier assumes that a harmless passenger jet flies to it and not an enemy bomber, or an insurance company assumes that 20-year-old potential clients have a 90% probability of dying next year. Thus, the insurance stops selling them life insurance policies. Extreme cases of faulty models sound frightening, though organizations quickly notice them. More dangerous are subtle manipulations. They are much harder to identify. How can the business detect a manipulated risk and pricing model that gives customers in southern Germany a 7% too low price while rejecting 10% of lucrative long-term customers in Austria? Such manipulations are often not apparent, but highly effective. They drive away loyal Austrian customers while generating direct financial losses in Germany due to unnecessary and unjustified discounts.

An external threat actor **manipulating** a specific AI model in the AI model repository or production systems sounds unlikely, at least for companies not in the focus of intelligence services. But even companies not in the spotlight of global cybercrime have to be aware of the threats of less-than-perfect AI models coming from the inside. Data scientists feeling mistreated (or being approached by competitors) can manipulate models much more effortlessly. Plus, even highly professional data scientists with goodwill make **mistakes**, resulting in mediocre or completely wrong models. Consequently, the internal quality assurance procedures for AI are crucial. AI-related quality assurance is often more in a highly agile startup mode and less rigid and sophisticated than for traditional software engineering.

The last remark already shifted the focus from manipulating the model to the overall creation and training process. **Lousy training data** implies bad models, no matter what quality metrics indicate. What would James Bond do if he does not want to be recognized by surveillance cameras? He would ingest a few hundred pictures of himself in a training set, tag himself as a bird, and retrain the AI model. From now on, surveillance cameras ignore him because he is a bird and not a burglar.

Training data mistakes or manipulations are severe and challenging to detect. They can happen at various places: in the original operational databases, in the data ingestion scripts, in the training data collection or data lake, and in the training area during the model training. If data scientists use external training and validation data sets, these are also potential ways for attackers. Who looks through millions of pictures to look for wrongly tagged James Bond images? Who understands why a command during test data preparation performs a left outer join and not a right outer join? Why are some values multiplied by 1.2 instead of 1.20001? And why does an upstream system discontinue sending negative values? Is it a mistake, sabotage, or improved data cleansing? Creating training data is highly complex, even without external threat actors.

All the threats and attacks discussed yet require access to training environments, production systems, AI models, or training data within an organization. Other threats do not require any of these. They succeed without touching any asset of the targeted organization.

Data scientists want to move fast and incorporate the newest algorithms of this quickly evolving field. They **download** pre-trained models, AI and statistics libraries, and publicly available datasets from the Internet. These downloads are the basis for building better company-specific models. At the same time, the downloads are a backdoor for attackers that can provide manipulated data or models, especially for niche models and training data.

Another attack type not requiring touching any assets are **adversarial attacks**. The idea is to manipulate input data so that a human eye does not detect the manipulation, but that the AI model produces the wrong result. For example, small changes in a picture or some new points on the street as illustrated in Figure 7-7 can result in the AI model not detecting a traffic light or getting a wrong understanding of whether to continue on the street to the right or whether to go full throttle straight and fall down the slope. Such attacks work only for AI models that are complex neural networks, not for simple linear functions.

Slowdown! Slowly to the right!   Full throttle, straight on!

Figure 7-7. *Understanding Adversarial Attacks – Attackers Change the Input Data Such as Pictures in a (Nearly) Invisible Way to Make the Neural Network Make Wrong Decisions. How Many Differences Do You See Between the Images?*

Availability Threats

The last threat scenario covers a potential non-availability. First of all, there is a risk of an **AI logic in production** or a **training area** outage. In the latter case, data scientists cannot work. If production systems are down, this impacts business processes. Non-availability is the consequence of attacks against **servers**. Typical attacks are crashing the applications with malformed input or flooding the server with (distributed) denial of service attacks.

An aspect with particular relevance for AI is the threat of **losing intermediate results or models**. In data science, processes tend to be more informal than in traditional software development regarding documentation and version management of artifacts. Such laxness can cause availability risks. For example, when a key engineer is sick, on holiday, or has left the company, his colleagues might not be able to retrain a model. Instead, they have to develop a completely new one, which takes a while and incurs costs.

From Threats to Risks and Mitigation

The difference between a conspiracist and a risk specialist is the consideration of probabilities. For example, the NSA could have manipulated this book by adding five sentences and removing an illustration with a dancing pink elephant from the previous page. Most probably, they can perform such an attack. However, why should they invest time and money in such an attack? Understanding the motivation and aims of threat actors and their technical and financial capabilities is crucial to address the highly relevant security risks and improve the overall security posture.

The confidentiality, integrity, and availability threats we discussed are generic. In an IT security risk assessment, risk assessors and AI experts take a closer look at which attacks are possible and rate the likelihood of attacks and their potential impact on the concrete AI environment. The result: a risk matrix as illustrated in Figure 7-8.

		Impact				
		Insignificant <25 TSD	**Minor** >25 TSD	**Moderate** >100 TSD	**Major** >1 Mio	**Catastrophic** > 10 Mio
Likelihood	**Almost certain** >1 per year	Moderate	High	High	Extreme	Extreme
	Likely Every year	Moderate	Moderate	High	High	Extreme
	Possible Every 5 years	Low	Moderate Ⓐ	Moderate	High	High
	Unlikely Every 20 years	Low	Low	Moder Ⓑ	Moderate	High
	Rare Even more seldom	Low	Low	Low	Moderate	Moderate

Ⓐ Unhappy data scientist steals training data   Ⓑ Activists identify unfair models

Figure 7-8. *Company-specific Risk Matrix for an AI Organization*

The sample risk matrix contains two threats. Risk A in Figure 7-8 reflects the possibility that an unhappy employee steals training data consisting of (anonymized) past customer shopping behavior. The financial impact rating is "minor," meaning between 50,000 and 250,000. The limited financial impact is due to its strong brand and unique logistics capabilities, making it challenging to copy the business model. The likelihood rating for such an attack is once every five years.

Based on the risk assessment and a risk matrix, the management has to decide which risks to accept and which to mitigate. **Mitigation** means investing in organizational measures or technical improvements. Later in this chapter, we discuss some more complex mitigation actions in detail. First, however, we introduce some typical mitigation actions in the context of AI for a better understanding already here. Examples are:

- Ensure that **access control and perimeter security** are in place for the training environment and production system. These measures reduce the risk of unwanted transfer of data or models to the outside and unwanted manipulations. In addition, they are essential if AI functionality resides in a public cloud.

- Enforce stringent quality processes and **quality gates**. The negligent data scientists are, in most cases, one of the highest risks for poor models. Four-eyes-checks and proper documentation of critical decisions reduce the risk of deploying inadequate models to production, though having a structured and comprehensive quality assurance process is even better.

- Ensure that external data, code libraries, and pretrained models come **from reliable sources**. Experimentations with such downloads are essential for data scientists. Still, some governance process is necessary when external downloads contribute to AI models deployed to production.

These three measures reduce the potential threat and attack surface of organizations. However, they do not make a proper risk assessment obsolete. A risk assessment takes a closer look at all threat scenarios and threats discussed above. AI is fun, and AI brings innovation. At the same time, organizations have to understand and manage their AI-related security risks. Defining mitigation actions only helps if there is clarity on who manages and addresses which risks. In the following, we take a closer look at potential actions.

Securing AI-Related Systems

For AI and analytics, data is what fuel is for your car: the more you have, the further and faster you can go. There is just one big difference: you burn the oil, but the data remains in your systems. An AI team piles up massive amounts of data within just a few months – a security nightmare. A scrupulous competitor has to turn only one data scientist against you – and she can extract and transfer most of your commercial data and intellectual property to the competitor. Consequently, this section has one ambition: elaborate on reducing this security risk without blocking the daily work of the data scientists and the AI organization.

Three traditional measures are the ticket to success together with three measures that require innovative thinking beyond standard IT security. The measures are:

1. Generic system hardening

2. Governance Processes

3. Compartmentalization

4. Advanced Techniques for Sensitive Attributes

5. Probing Detection

6. Cloud-AI Risk Mitigation

AI and security specialists apply them to the various AI components as Figure 7-9 illustrates.

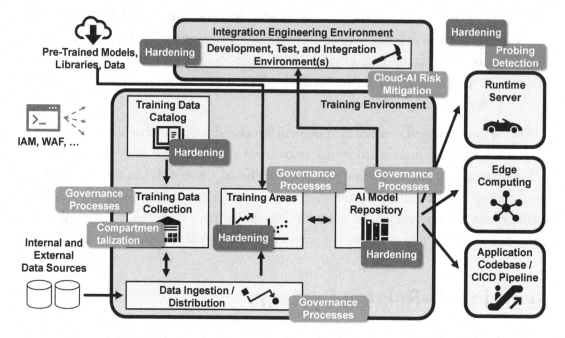

Figure 7-9. *Security Measures for AI Organizations' Application Landscapes*

System Hardening

The generic system hardening covers the traditional IT security measures such as network zones and firewalls, integrating security services such as Internet and access management (IAM) solutions or web application firewalls (WAF), and configuring systems securely, that is, restricting IP ranges.

These are standard activities known from "normal" software. They apply as well for any AI system or component, be it a training area, repositories, data lakes, or anything else. The aim: making sure that unauthorized persons do not get access to the systems plus enabling data scientists and engineers to perform their work.

Governance

The distinction between adequate and necessary vs. unneeded and risky access to training data requires human judgment. The outcome of the judgment should be transparent and repeatable. Similar situations should result in similar decisions and requesters want to understand who is involved in the decision process and how far the decision-making proceeded. Governance processes help – and AI organizations benefit from setting up two especially relevant ones for data uploads and for data usage (Figure 7-10). The data upload governance process verifies whether the data loaded into the training environment is adequate. The usage governance process verifies whether the data can be used for a particular use case.

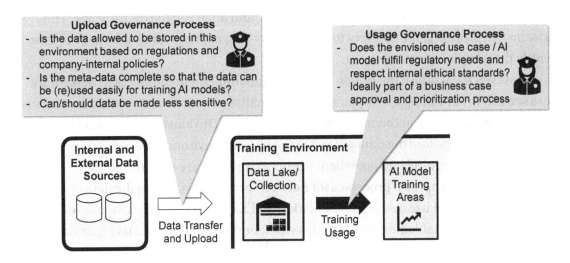

Figure 7-10. *Governance Processes for AI Training Environments*

The separation of the two processes is a cost- and throughput-time optimization. So, a first project needing telemetry data from a new turbine implements the data transfer and preparation and goes through the upload approval process. Ideally, ten or twenty later projects can reuse the data. They only need approval for using the data for their purposes. They can skip the upload approval and reuse the implemented and approved data copy and preparation infrastructure, saving time and money.

The **data upload governance process** looks, from an approval perspective, whether data can and should be copied into an AI training environment. Data owners, data privacy officers, legal teams, and IT security typically drive the decision-making. Obvious concerns relate to data privacy when AI training environments are in a different jurisdiction than the humans whose data the company analyzes. A second concern relates to companies in highly competitive and sensitive industry sectors or governmental agencies. They might hesitate to put every piece of information in every public cloud, no matter how good its AI features are.

While the data upload governance process has a control function, it also helps make training data more valuable by easing (re-)use. This governance process can act as a checkpoint for documentation in the data catalog. It can enforce that data is only put into the training environment when the metadata is complete. This means: Data scientists must describe the content of their data, have assessed and documented the data quality, clarified potential usage restrictions, and analyzed the data lineage. This governance process is the shortcut to high-quality data catalogs for all data sets, at least within the AI training environment.

However, there is also the option to discuss and implement measures to **make data less sensitive**. For example, shopping cart information related to individual customers is more problematic than anonymous shopping cart data. Obviously, the natural point for anonymization is before the data upload to the training environment.

Ideally, the usage-related governance process is part of the organization's management and approval processes for new (AI) projects. Following the data upload and before an actual use case implementation and AI model training, it focuses on approvals, not on enablement. It questions whether the intended data usage or the envisioned AI model violates data privacy laws, internal ethics guidelines, or any other policy or regulation. Again, data owners, data privacy officers, or legal and compliance teams are the natural decision-makers.

Data Compartmentalization and Access Management

Compartmentalization balances two contrary wishes: the wish for fine-granular data access control plus the wish and need for manageability. Achieving both at the same time is possible for **repositories** and the AI model training areas. If a specialist works on a project, he gets access to the specific training data; otherwise, not. Also, figuring out which application has to get models from a particular repository is straightforward. The challenge is the access management for data lakes or the **data collection** of the AI organization within their AI model training environment.

Access to all data for every data scientist is not an option. Neither is managing data access separately for thousands of data sets and data classes. The latter might be a surprise. Super fine-granular access rights appeal, at least at first glance. However, data scientists and data owners cannot work with them on a daily basis. The complexity is too high for the human mind. As a result, they would stop following the need-to-know principle to prevent the AI organization from being blocked. They move to a "grant access that is not obviously unnecessary" model, approving much more data access than needed. So, super fine granular access control looks good on paper, but fails in reality.

Suppose a data owner from the business knows his data and the data model very well. He can handle ten or twenty subsets, not one hundred and not thousands. AI organizations need a compartmentalization approach to limit the number of data access control roles and adjust them to their organization's actual size and complexity. A starting point is following a three-dimensional compartmentalization approach (Figure 7-11). It works for medium enterprises and scales even to the largest corporations of the world.

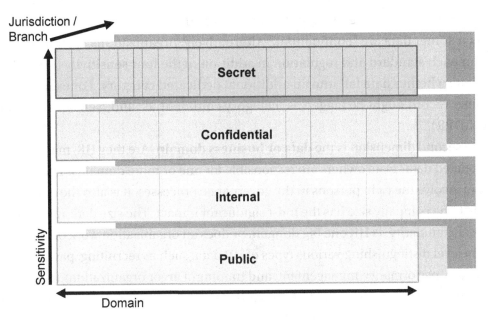

Figure 7-11. *The Three Dimensions of Compartmentalization*

The first dimension is **sensitivity**, typically categorized based on four levels: public, internal, confidential, and secret. Public data is (or could be made) available on the web, for example, statistics published by public health organizations or product descriptions in marketing brochures. Then, there is internal data. It means every employee (with a good reason) can access it. Customer contact information is one example. So is a categorization whether a customer is a top or a standard customer, or a customer frequently causing trouble. Complex offerings in a B2B context or a WHO tender are examples of confidential data. Deal or no deal can have a significant impact on the company's bottom line. And any potential disclosure of the offer to competitors might result in losing the deal for sure. Finally, there is secret data. Passwords or (master) keys for encryption and signing files belong to this category. Information potentially impacting a corporation's stock exchange price might – before disclosure – fall into this category. The latter examples also illustrate that data sensitivity can change over time. For example, once such stock-price relevant information is out, it becomes public data.

Additional standards and regulations impact the data sensitivity classifications in certain industry sectors. Examples are the already mentioned GDPR, Health Insurance Portability and Accountability Act (HIPAA), or Payment Card Industry Data Security Standard (PCI-DSS). Organizations can model the impact in different ways. The regulatory standards can influence to which of the four categories data belong. For example, Singapore and Channel Island bank customer data are "secret," other customer data are "confidential." Alternatively, organizations can introduce flags for each standard and regulation in addition to the four sensitivity levels. These flags state whether data fall into an additional dedicated category. For example, a customer record might be tagged as "category confidential" and "sensitive personal data (GDPR)."

The second dimension is the **data or business domain**. Are they HR, marketing, or R&D-related data? Data owners are responsible for one or more domains. The domain helps to involve the right persons in the governance processes. It is also the dimension on which the company size has the most significant impact. The size determines the approval granularity. Is HR one big domain, or does an organization work on a sub-domain level distinguishing various types of HR data, such as recruiting, payroll, insurances, performance management, and training? Larger organizations tend to have a more granular approval level. They have more staff to handle requests, more sensitive data, and a more complex data model.

The third dimension reflects the company's **organizational structure and various jurisdictions** in which the company operates and to which data might belong. Different subsidiaries might have different internal rules. Different jurisdictions regulate data storage and usage and artificial intelligence differently.

To conclude, domains, organizational structures, and jurisdictions are natural ways to divide work, manage access rights, route requests in governance processes, and determine quickly which rules and regulations apply. The data sensitivity dimension allows for assessing quickly whether a request requires a more profound analysis or can take a shortcut.

Advanced Techniques for Sensitive Attributes

Certain data types stand out from the rest of the training data, because they pose a high risk if lost or exposed. First, these are data items allowing to identify individuals, such as Social Security Numbers. Second, there are sensitive personal and protected classes such as union membership or race. Here, it makes sense to consider different approaches beyond typical access control. The options are:

- Not copy them to the AI environment. In case protected classes and sensitive personal data must not be used for AI model training, why copy such data? The same applies to identifiers. If not needed, one can remove them when importing the data into the AI environment as training data.

- Anonymizing data means making it completely impossible to identify individuals to which data belongs, even not using additional information and tables. If you replace names with "XXXX," the result is anonymous data (at least if there are no other columns, e.g., with IDs). AI organizations have really to check in detail in case anonymization is mandatory. Often, this term is also used for the third option, veiling, as well.

- Veiling data, that is, making it more challenging and time-consuming to determine to which individual the data belong. It reflects situations where, for example, a table contains synthetic names and account numbers, but in the end, there is still a table for mapping synthetic names and numbers to the real ones, for example, to ease fixing issues.

Probing Detection

Partners, competitors, or customers: these are potential attackers trying to reengineer or misguide automated ratings or decision making. A classic bank with brick branch offices might want to understand the pricing model of an online bank to target customers not getting "good" online offers with high-margin credits (Figure 7-12, left). A car repair shop might want to understand which kind of picture of a damaged car lets the insurance pay the highest compensation for the repairs. Is it a long shot or a detailed picture, should it be with intense colors – or are black and white images more profitable (Figure 7-12, right)?

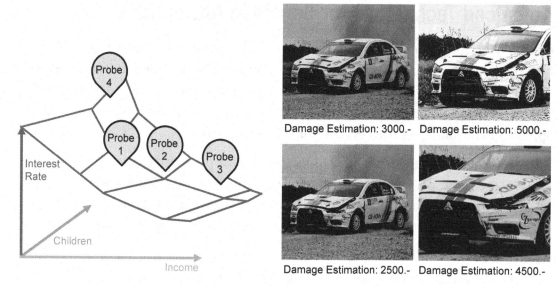

Figure 7-12. *Probing for Model Reconstruction (Left) and for Searching for Weaknesses or Sweet-Spots (Right), Picture Source: Pixabay*

If attackers cannot directly access an AI model, they can – if exposed to the Internet – probe it and send many requests with different value combinations such as income, property, age, child support, or other parameters for our bank example. The attackers collect the replies and use the data to train a copycat AI model behaving similarly to the bank's original one. The attackers have no access to the original model, but they can analyze their copycat AI model instead. Thus, AI organizations and IT security specialists have to prevent the probing of an AI model – without disturbing "normal" customers' requests. Potential approaches are:

– Restrict the number of requests per hour for each IP address – or even confuse attackers with wrong results.

– Restrict anonymous access to the systems and require registration before providing any estimations.

– Use Captchas to prevent at least automated probing.

– Do not provide the results of the estimation directly online, but only per email, letter, or phone call.

Companies have to balance the risk of losing potential customers that might find, for example, the process too inconvenient of having to wait for a letter against the risk that competitors or partners probe the model.

To conclude: Companies exposing AI models to the public or to partners, for example as part of the business process, should monitor the processes and requests for anomalies and hints for fraud or espionage.

Cloud-AI Risk Mitigation

Many AI organizations are ahead of their time when it comes to moving to the cloud, which brings them into trouble. Massive amounts of data, highly fluctuating compute needs, and the availability of sophisticated AI frameworks in the public cloud foster a quick cloud adoption by data scientists and AI specialists. The rest of the company and the IT security organization might be on the journey to the cloud. Most AI organizations are already there. As a consequence, AI organizations have to take over more security-related tasks compared with traditional on-premise environments. Thus, AI managers, data scientists, and engineers benefit from a basic understanding of the main security risks.

Before elaborating on the risks of being *in* the cloud, we look at the unique **compliance risks** when moving *to* the cloud, especially to the big global players. First, public cloud providers offer many different services, from compute and storage to very specific niche services for IoT or the blockchain. Not every service is available in each region – and even if the actual configuration of the data scientists and engineers determine whether the data end up in a data center in an adequate **jurisdiction**. One wrong click during configuration and data scientists or data engineers transfer EU data to the US, often a big issue. Second, the **Cloud Act** means that US authorities might get access even to data in non-US data centers. These data transfers or the potential data exposure can be a no-go for regulators. Third, there is the **sanctions** risk. In such a case, a company might have to switch to a new provider in a different juristiction immediately. In an on-premise world, the software continues to run when a company is put under sanctions, even if security patches are withheld. In contrast, access to all data and software is blocked immediately in the cloud – a real risk even for smaller companies and governmental offices in OECD countries as Sassnitz harbor in Germany exemplifies. The port handles 2.2 Mio tons of freight per year, nothing compared to 469 Mio in Rotterdam. It is owned by the local city and a German federal state. Economically irrelevant, solid owners – and in the middle of a pipeline controversy resulting in the US threatening to sanction the port. What would happen with such a company if its logistics, for example, depend on AI frameworks in the public cloud?

Once the AI organization decides to move to the cloud, particular security risks emerge and become newly relevant. Figure 7-13 provides an overview of the risk areas requiring cloud-specific **technical security measures**. They are important if the AI organization has an extra cloud or a separate tenant, but only to a lesser degree if running in the same tenant as the rest of the company.

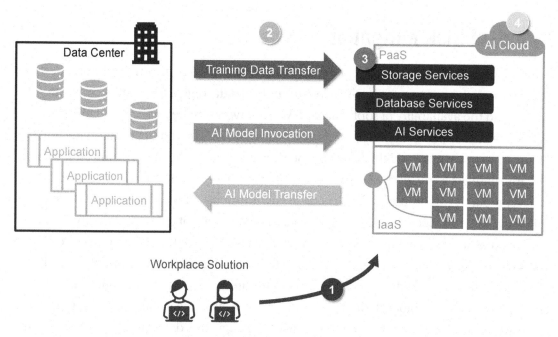

Figure 7-13. *Cloud Security Risks for AI Organizations*

The first risk area relates to user access. Data scientists and engineers have to connect and log in to the cloud (Figure 7-10, 1). If the AI organization does not use the same cloud environment as the rest of the organization, they might bypass the company's central **user and access management** and Active Directory. So, managing new employees or dealing with leavers or ones moving to new positions becomes an issue. There is one thing more inconvenient than having an employee guided out of the building by a security guard: if this former employee has access to the cloud for the next weeks and months allowing for revanche because the Active Directory is locked, but not a cloud account "in the wild." Furthermore, most companies enforce **multi-factor authentication**. Logging in just with a user name and password is frowned upon these days. Employees need RSA keys or special apps on their mobile phones. The public clouds provide these features as well, but they are not enforced by default.

Service accounts are the second area, quite a similar topic (Figure 7-10, 2). The on-premise world or the other company's cloud environments have to interact with the AI organization's cloud tenant:

- Training data must get to the AI environment.

- A trained AI model might have to be moved to the on-premise world if the runtime environment is there or if the model becomes part of the application code.

- A trained model might rely on the cloud as an AI runtime environment. Then, various applications of the company have to be able to invoke the AI models on the cloud for interference.

The challenge with service accounts is preventing issues with stolen passwords or access keys. It would be a catastrophe if a data engineer leaves the company, copies a password, and continues to have access to the cloud for weeks and months.

Internet-exposed platform-as-a-service services are the third risk area (Figure 7-10, 3) – and they are unique to public clouds. Databases or object storage are web services. Applications – and attackers – can reach them potentially from the public Internet. Both can invoke AI models and many more services and resources in the same way. Engineers should configure them such that only authenticated users coming from specific IP addresses have access. However, misconfigurations can result in public exposure. The majority of cloud data leakages were the result of such misconfigurations, not of sophisticated attacks.

For clarification, the risk is not so prominent for infrastructure-as-a-service services such as VMs. Accidentally opening up a VM to the Internet requires many more misconfigurations than in the PaaS world.

The fourth topic is **security operations** in the cloud. Who takes care of security incidents, inspects logs, and analyzes warnings? (Figure 7-10, 4)

The cloud-risk areas are not specific to AI organizations. Every engineer should be aware of them. However, at the moment, most companies are in a transitional phase. AI organizations might not have the usual support from IT security. Suppose the AI organization decides unilaterally to set up an AI cloud. In that case, they are responsible for security engineering and security operations. They cannot assume that the IT security organization helps them. AI managers might realize quickly: auditors do not sympathize with an AI cloud with large amounts of data and insufficient IT security know-how. Finally, to make things worse, the auditors expect that the AI cloud follows the same norms and standards as the rest of the organization, including, for example, the ISO 27000 information security standard.

The ISO 27000 Information Security Standard

Standards are not a creative's – or an IT specialist's – darlings. Most engineers and data scientists prefer working in sectors without heavy regulations and strict norms. Even if there are engineering or organizational standards such as Scrum, SAFe, Cobit, ITIL, or TOGAF, they allow heavy tailoring and interpretation. One broadly relevant exception to the rule of ease is the ISO 27000 standard. If the senior management decides to take this norm seriously, the IT security department and auditors get persistent. IT norms did not impact traditional statisticians' teams. The situation differs for AI organizations and data scientists writing code, providing web services, and running and hosting software. They fall under the same rules as any software engineering and systems integration team.

The ISO 27000 standard is a **standard family** consisting of a variety of documents with different purposes (Figure 7-14). ISO 27000 is the vocabulary standard. It provides a basic overview and defines the terminology. Then, there are binding standard documents: ISO 27001 and ISO 27006. ISO 27006 is irrelevant for AI teams since it addresses auditing organizations only. In contrast, ISO 27001 is the bible for any certification. It states all certification requirements. The standard has a main section and appendix A. The main section defines a general information security framework. Topics such as top management involvement or the need for an incident management system are more relevant for senior managers and IT security departments than for AI teams and software engineers. Appendix A lists concrete security topics ("controls") to be implemented. Some apply directly to AI organizations, including the needs for:

- Identifying and classifying assets such as code, data, or documentation and defining access control mechanisms

- Securing development environments and test data

- Performing acceptance and security tests

- Change control procedures in development and production to prevent unwanted changes and clear rules who can do which installations

ISO 27001 compliant companies often translate these requirements into internal guidelines and directives, such that many organizations implement the requirements without engineers and AI specialists being aware.

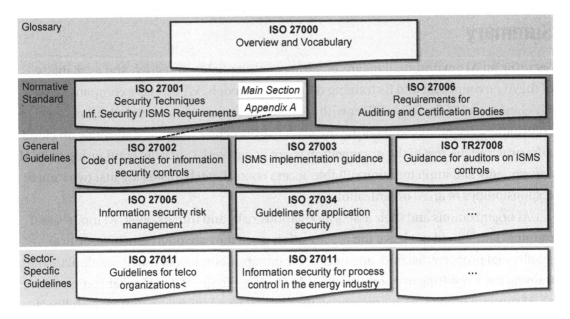

Figure 7-14. The ISO 27000 Norm Family

As mentioned, the ISO 27000 standard family comprises many more norms and documents. Though the ISO 27001 standard is the only normative binding document for IT organizations, ISO 27002 is worth taking a closer look at since it helps to set up the controls of appendix A of ISO 27001.

Additional guidelines focusing on specific aspects. ISO 27003, for example, looks at information security management systems, and ISO 27005 at risk management. Sector-specific guideline standards are also available, for example, for the financial services industry (now withdrawn), telcos, or the energy sector. However, most companies focus on ISO 27001/27002 and do not pay much attention to the additional documents.

A number behind the standard reflects the publishing year and allows to distinguish different versions, for example, the ISO27001:2013 version from the older version, ISO27001:2005. At the moment, there is a peculiar situation with a new draft version of the ISO 27002 standard out, but not a new ISO 27001 draft, which would be the actual binding norm. However, at this moment, there seems not to be much change impacting AI organizations.

For most AI organizations, it is more an **emotional shock** than a big issue to adhere to ISO 27001, at least if they are in contact and aligned with the IT security organization beforehand.

Summary

Securing an AI environment means ensuring confidentiality, integrity, and availability of the AI environment and its training data and AI models. While every company has a Chief Information Security Officer with a smaller or larger team, AI organizations have to contribute substantially to securing their AI environment, especially if they rely on cloud infrastructure the rest of the organization is not using (yet). Correctly configuring their AI components and implementing suitable access control mechanisms are just two sample responsibilities of an AI organization.

AI organizations and their assets face various risks and threats. First, AI models and training data should not leave the organization since they are often sensitive, critical intellectual property. Second, an attacker (or unhappy employee) might manipulate the training data, resulting in poorly working AI models. Finally, if the AI organization runs an AI runtime server which executes all the company's AI models, a non-availability of the AI runtime server impacts all processes somehow depending on requesting AI model inference services on the non-available server.

As a consequence, AI organizations should conduct a risk assessment together with the security organization to identify weaknesses and implement technical and procedural best practices for securing their AI environment and their assets.

CHAPTER 8

Looking Forward

It is a characteristic of excellent AI line managers and project managers to direct world-class data scientists to build AI solutions that match the company's strategic business goals. The challenge is not having ideas on how AI could change the world. The challenge is to build practical, helpful AI solutions delivering concrete benefits and, afterward, running these AI solutions over the years.

The various book chapters elaborated all relevant success factors from shaping and managing AI projects, organizing testing and quality assurance, integrating AI components into broader application landscapes and operating them, to incorporating ethics, compliance, and security-related topics. These are the concepts every AI line or project manager needs to set up, organize, and manage an AI department and its project portfolio.

When data scientists move to a new project, they might require a deep understanding of different AI algorithms or frameworks. In contrast, AI managers can master various projects with the know-how presented in this book. Their tasks are very similar, whether the projects are about voice biometrics, text understanding, support chatbots, automated credit ratings, or autonomous cars and planes. Thus, if you are an AI project or line manager and made it to this page, it is time for me to congratulate you for taking the time and investing so much effort. I hope you enjoyed reading the book as much as I enjoyed structuring the topic and bringing it to paper. I wish you many successful AI projects in the years to come. Plus, I am sure you can move your company's AI organization to the next level with the ideas you learned from this book!

© Klaus Haller 2022
K. Haller, *Managing AI in the Enterprise*, https://doi.org/10.1007/978-1-4842-7824-6_8

Index

A

Access control, 192, *See also* Identity and
 access management (IAM)
Access management, 196, *See also* Identity
 and access management (IAM)
Accountability principle, 93
Accuracy, 64, 69, 72, 93
Adjusted R2 Metric, 68
Adversarial attacks, 190
AI act, 94
 high-risk AI systems, 95
AI-driven solutions, 81
AI environments, 149
 Application Landscape, 150
 Data Catalog, 160–163
 Executing Models, 164
 Ingestion, 150
 model and code repositories, 163
AI ethics governance, 90, *See also*
 Governance
AI frameworks, 24, 56
 AI models, 47, 88, 185, 187
 Ethics, 88
AI models, 47, 88, 185, 187
AI operations services, 111, 128, 140
AI organizations, 107, 114, 125, 128,
 140, 168
AI project deliverables, 16
AI projects, 110
AI project services, 118, 140
AI regulations, 92, *See also* Regulations

AI runtime server, 59, 82, 185, 187
AI service organization, 131
AI services, 55, 108, 112, *See also* AI
 service types
 human factor, 109
AI services layer, 25
AI service types, 110, 112
 service boutique, 113
 service factory, 114
 service mall, 114
 service shop, 114
AI translator, 28, 67, 119
Amazon Comprehend Medical, 26
Anonymizing data, 199
Ansoff Matrix, 13
Anti-virus software, 178, *See also* Malware
Application management (AM), 129
Application Specific Integrated Circuits
 (ASICs), 24
Architecture, 166, 172, 194
Artificial general intelligence, 4
Artificial intelligence Act, *see* AI act
Artificial narrow intelligence, 4
Artificial superintelligence, 4
Assets, 185
Associations, 31
Audit logs, 140
Audits, 93
Auto machine learning (AutoML), 55
Automated model scoring, 136
Autonomous algorithms, 5

© Klaus Haller 2022
K. Haller, *Managing AI in the Enterprise*, https://doi.org/10.1007/978-1-4842-7824-6

Printed in the United States
by Baker & Taylor Publisher Services